An Introduction to Microelectromechanical Systems Engineering

For a listing of recent titles in the *Artech House Microelectromechanical Systems Library,* turn to the back of this book.

An Introduction to Microelectromechanical Systems Engineering

Nadim Maluf

Artech House
Boston • London

Library of Congress Cataloging-in-Publication Data
Maluf, Nadim.
   An introduction to microelectricomechanical systems enginnering / Nadim Maluf.
     p.   cm. — (Artech House MEMS library)
   Includes bibliographical references and index.
   ISBN 0-89006-581-0 (alk. paper)
   1. Microelectromechanical systems.   I. Title.   II. Series.
TK7875 .M35   2000
621.381—dc21   99-052406
                                             CIP

British Library Cataloguing in Publication Data
Maluf, Nadim
   An introduction to microelectromechanical systems —
(Artech House MEMS library)
   1. Microelectromechanical systems
   I. Title
   621.3'81

   ISBN   0-89006-581-0

Cover design by Lynda Fishbourne.   Cover illustration by Jerry Gist.

© 2000 ARTECH HOUSE, INC.
685 Canton Street
Norwood, MA 02062

All rights reserved. Printed and bound in the United States of America. No part of this book may be reproduced or utilized in any form or by any means, electronic or mechanical, including photocopying, recording, or by any information storage and retrieval system, without permission in writing from the publisher.
   All terms mentioned in this book that are known to be trademarks or service marks have been appropriately capitalized. Artech House cannot attest to the accuracy of this information. Use of a term in this book should not be regarded as affecting the validity of any trademark or service mark.

International Standard Book Number: 0-89006-581-0
Library of Congress Catalog Card Number: 99-18034

In memory of my father-in-law, Gabriel Amin Hoché, and the passengers and crew members who perished on September 2, 1998, on board Swissair flight 111. May their memory evoke in engineers and technologists a continuous striving for absolute safety.

To my parents, Samira and Elias Maalouf

Contents

Foreword	*xiii*
Preface	*xvii*

[1] MEMS: A Technology from Lilliput — 1

The promise of technology	1
What are MEMS—or MST?	3
What is micromachining?	6
Applications and markets	6
To MEMS or not to MEMS?	7
Standards	9
The psychological barrier	10
Journals, conferences, and Web sites	10
List of journals and magazines	*11*
List of conferences and meetings	*12*
Summary	13
References	13

2 The Sandbox: Materials for MEMS — 15

Silicon material system — 16
- Silicon — 16
- Silicon oxide and nitride — 23
- Thin metal films — 23
- Polymers — 25

Other materials and substrates — 25
- Glass and quartz substrates — 26
- Silicon carbide and diamond — 26
- Gallium arsenide and other group III-V compound semiconductors — 27
- Shape-memory alloys — 27

Important material properties and physical effects — 28
- Piezoresistivity — 29
- Piezoelectricity — 31
- Thermoelectricity — 35

Summary — 37

References — 37

3 The Toolbox: Processes for Micromachining — 41

Basic process tools — 42
- Epitaxy — 43
- Oxidation — 44
- Sputter deposition — 44
- Evaporation — 45
- Chemical vapor deposition — 46
- Spin-on methods — 50
- Lithography — 51
- Etching — 55

Advanced process tools	70
Anodic bonding	70
Silicon-fusion bonding	71
Grinding, polishing, and chemomechanical polishing (CMP)	72
Sol-gel deposition methods	74
Electroplating and molding	75
Combining the tools—examples of commercial processes	75
Polysilicon surface micromachining	77
Combining silicon fusion bonding with reactive ion etching (SFB-DRIE)	79
SCREAM	81
Summary	82
References	83

4   The Gearbox: Commercial MEM Structures and Systems   87

General design methodology	88
Techniques for sensing and actuation	90
Common sensing methods	90
Common actuation methods	91
Passive MEM structures	95
Fluid nozzles	95
Inkjet print nozzles	97
Sensors	99
Pressure sensors	99
High-temperature pressure sensors	104
Mass flow sensors	105
Acceleration sensors	108
Angular rate sensors and gyroscopes	119
Radiation sensors—infrared imager	134

	Carbon monoxide gas sensor	*136*
	Micromachined microphone	*138*
	Actuators	142
	Digital Micromirror Device™	*142*
	Micromachined valves	*147*
	Summary	156
	References	157
5	**The New Gearbox: A Peek Into the Future**	**161**
	Passive micromechanical structures	162
	Hinge mechanisms	*162*
	Sensors and analysis systems	163
	Miniature biochemical reaction chambers	*163*
	Electrophoresis on a chip	*168*
	Microelectrode arrays	*171*
	Actuators and actuated systems	176
	Micromechanical resonators	*176*
	High-frequency filters	*180*
	"Grating light valve" display	*183*
	Optical switches	*187*
	Micropumps	*190*
	Thermomechanical data storage	*192*
	RF switch over gallium arsenide	*197*
	Summary	198
	References	198
6	**The Box: Packaging for MEMS**	**201**
	Key design and packaging considerations	202
	Wafer or wafer-stack thickness	*204*
	Wafer dicing concerns	*204*

Thermal management	*205*
Stress isolation	*207*
Protective coatings and media isolation	*208*
Hermetic packaging	*210*
Calibration and compensation	*211*
Die-attach processes	212
Wiring and interconnects	216
Electrical interconnects	*216*
Microfluidic interconnects	*220*
Types of packaging solutions	222
Ceramic packaging	*223*
Metal packaging	*228*
Molded plastic packaging	*230*
Summary	235
References	235
Glossary	**239**
About the Author	**251**
Index	**253**

Foreword

According to my best recollection, the acronym for Microelectromechanical Systems, *MEMS*, was officially adopted by a group of about 80 zealots at a crowded meeting in Salt Lake City in 1989 called the Micro-Tele-Operated Robotics Workshop. I was there to present a paper that claimed MEMS should be used to fabricate resonant structures for the purposes of timekeeping, and I was privileged to be part of this group of visionaries for one and a half exciting days. (The proceedings may not be in print any longer; however, I recall they were given in IEEE Catalog #89TH0249-3.)

Discussion at the workshop about the name of this new field of research raged for over an hour, and several acronyms were offered, debated, and defeated. When the dust settled, I recall that Professor Roger Howe of the University of California at Berkeley stood up and announced, "Well, then, the name is MEMS." In this way, the group came to a consensus. The research they conducted, unique among any that was being conducted in the United States (or the world for that matter), would thereafter be known as "MEMS."

In those early, heady, exciting, and terribly uncertain days, those in the nascent field faced many issues that researchers today would find hard to remember. For example, our hearty band constantly worried if any scholarly journal would publish the papers we wrote. Sources of research funding were hard to find and difficult to maintain. MEMS fabrication was itself a major issue, and the frequent topic of conversation was about the nature, properties, and standardization of the polysilicon that the pioneering researchers were using to demonstrate the early, elementary structures of the day. Even the most daring and idealistic of students occasionally turned down an offer to work with the faculty of that era. The work appeared too far-fetched for the taste of even the green-eyed zealots among the graduate student population.

In the ten years that have passed since the momentous events of that watershed workshop, the National Science Foundation (NSF) has funded a set of MEMS projects under its "Emerging Technologies Initiative," headed by George Hazelrigg. NSF funding continues to this day. The Defense Advanced Projects Research Agency (DARPA) put nearly $200 million into MEMS research. Numerous MEMS journals have sprung up, and the rate of filing of MEMS patents had reached over 160 per calendar year in 1997. The skeptics that predicted the collapse of the field in 1990 are now confronted with the fact that, in 1997, there were 80 U. S. companies in the MEMS field. The combined total world market of MEMS reached approximately $2 billion. In addition, the most conservative market studies predict a world MEMS market in excess of $8 billion in 2003. In a phrase, MEMS has arrived. Despite all the rosy news, there remain significant challenges to face in the MEMS field. One of these I call the challenge of the "500 MEMS Companies" and the other, the "10,000 MEMS Designers." For the field to take full root and become ubiquitous there must be an unprecedented training of tens of thousands of MEMS engineers. Already, the demand for MEMS experts has far outstripped the ability of academia to train them. The only hope is for existing engineers to learn the basics of MEMS and then go up the MEMS learning curve in the traditional way, i.e., learning by doing.

Here is where this book plays an essential role on the national stage. Dr. Nadim Maluf has put together one of the finest MEMS primers that you can find on the bookshelf today. Written in a no-nonsense, clear style, the book brings the practicing engineer and student alike to an understanding of how MEMS are designed and fabricated. Dr. Maluf's book concentrates mostly on how to design and manufacture MEMS. This is to be expected of Dr. Maluf, who has impeccable MEMS credentials. Trained in MEMS for his Ph.D. at Stanford University, Dr. Maluf has spent his post-doctoral career as a practicing MEMS engineer and manager at Lucas NovaSensor, one of the early MEMS companies. His

Foreword

industrial career has focused both on bringing MEMS products successfully to market, and on defending his company's market share against encroachment by other technologies. Since this book is written from Dr. Maluf's practical perspective, it is sure to have lasting value to the myriad of engineers and executives who are struggling to find a way into the field of MEMS. This book will also serve as a useful resource for those already in the field who wish to broaden their expertise in MEMS fabrication. When I reviewed the manuscript, I was ready to offer Dr. Maluf a great deal of suggestions and corrections. I was quite humbled to realize that, instead, I was eager to have a copy of the new book on my own bookshelf. It will serve as a reference not only for myself, but also for the students and engineers who frequently ask me, "What book should I buy to learn how to make MEMS?"

Albert ("Al") P. Pisano, Ph.D.
MEMS Program Manager
DARPA
June 1999

Preface

A few years ago I stood before an audience at a customer's facility explaining the merits of micromachining technology. The small conference room was packed, and all ears were attentive. Everyone was eager to learn about this mysterious buzzword, "MEMS." Although many in the audience were nodding in a sign of comprehension, the glazed looks on their faces betrayed them. This experience is not unique, but one that is repeated frequently in auditoriums around the world. The technology is simply too broad to be explained in a short lecture. Many technical managers, engineers, scientists, and even engineering students with little or no previous experience in microelectromechanical systems are showing keen interest in learning about this emerging technology. This book is written for those individuals.

In this book I sought to introduce the technology by describing basic fabrication processes and select examples of devices and microsystems that are either commercially available, or show great promise of becoming products in the near future—practical examples from the "real world." The objective is to provide a set of representative cases that can give the reader a global understanding of the technology's foundations, and a sense of its diversity. The text describes the basic operation and fabrication of many devices, along with packaging requirements. Inspired by the adage "a picture is worth a thousand words," I have included numerous descriptive schematic illustrations. It is my hope that scanning these illustrations will aid the reader in quickly developing a basic familiarity with the technology. Suggestions at the end of each chapter for additional reading and an extensive glossary will supplement the main text.

The following is an overview of each chapter in the book.

Chapter 1—MEMS: A Technology from Lilliput. This introductory chapter defines the scope of the technology and the applications it addresses. A short analysis of existing markets and future opportunities is also included.

Chapter 2—The Sandbox: Materials for MEMS. This chapter reviews the properties of materials common in micromachining. The emphasis is on silicon and materials that can be readily deposited as thin films on silicon substrates. Three physical effects, piezoresistivity, piezoelectricity, and thermoelectricity, are described in some detail.

Chapter 3—The Toolbox: Processes for Micromachining. Various fabrication techniques used in semiconductor manufacturing and micromachining are introduced. These include a number of deposition and etch methods, and lithography. The discussion on etch methods covers the topics of anisotropic etching, dependence on crystallographic planes, and deep-reactive-ion-etching. Three complete manufacturing process flows are described at the end.

Chapter 4—The Gearbox: Commercial MEM Structures and Systems. This chapter includes descriptions of a select list of commercially available micromachined sensors and actuators. The discussion includes the basic principle of operation and a corresponding fabrication process for each device. Among the devices are pressure and inertial sensors, a microphone, a gas sensor, valves, an infrared imager, and a projection display system.

Chapter 5—The New Gearbox: A Peek into the Future. The discussion in this chapter centers on devices and systems still under development, but with significant potential for the future. These include biochemical and genetic analysis systems, high frequency components, display elements, pumps, and optical switches.

Chapter 6—The Box: Packaging for MEMS. The diverse packaging requirements for MEMS are reviewed in this chapter. The basic techniques of packaging sensors and actuators are also introduced. A few nonproprietary packaging solutions are described.

The writing of a book usually relies on the support and encouragement of colleagues, friends, and family members. This book is no exception. I am grateful to Al Pisano for his general support and for recognizing the value of an introductory book on MEMS. I would like to thank Greg Kovacs, Kirt Williams, and Denise Salles for reading the manuscript and providing valuable feedback. They left an indelible mark of friendship on the pages of the book. I am thankful to many others for their comments, words of encouragement, and contributions. To Bert van Drieënhuizen, Dominik Jaeggi, Bonnie Gray, Jitendra Mohan, John Pendergrass, Dale Gee, Tony Flannery, Dave Borkholder, Sandy Plewa, Andy McQuarrie, Luis Mejia, Stefani Yee, Viki Williams, and the staff at NovaSensor, I say "Thank you!" For those I inadvertently forgot to mention, please forgive me. I am also grateful to DARPA for providing partial funding under contract N66001-96-C-8631. Last but not least, words cannot duly express my gratitude and love to my wife, Tanya. She taught me, over the course of writing this book, the true meaning of love, patience, dedication, understanding, and support. I set out in this book to teach technology, but I finished learning from her about life.

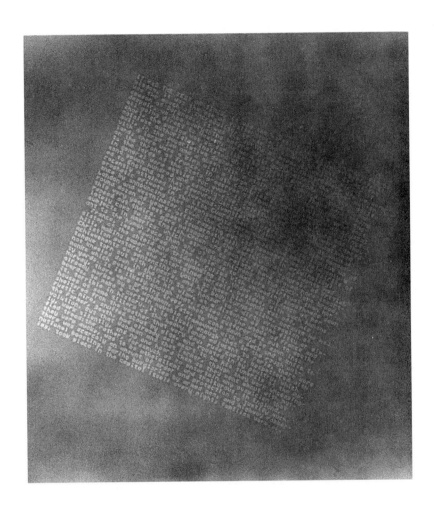

"It was the best of times, it was the worst of times, it was the age of wisdom, it was the age of foolishness ... " from *A Tale of Two Cities* by Charles Dickens, engraved on a thin silicon nitride membrane. The entire page measures a mere 5.9 μm on a side, sufficiently small that 60,000 pages—equivalent to the Encyclopedia Britannica—can fit on a pinhead. The work, by T. Newman and R.F.W. Pease of Stanford University, won the Feynman challenge in 1985. Courtesy of *Engineering & Science Magazine*, CalTech.

CHAPTER 1

Contents

The promise of technology

What are MEMS—or MST?

What is micromachining?

Applications and markets

To MEMS or not to MEMS?

Standards

The psychological barrier

Journals, conferences, and Web sites

Summary

MEMS: A Technology from Lilliput

… And I think to myself, what a wonderful world … oh yeah!

Louis Armstrong

The promise of technology

The ambulance sped down the Denver highway carrying Mr. Rosnes Avon to the hospital. The flashing lights illuminated the darkness of the night, and the siren alerted those drivers who braved the icy cold weather. Mrs. Avon's voice was clearly shaking as she placed the emergency telephone call a few minutes earlier. Her husband was complaining of severe heart palpitations and shortness of breath. She sat next to him in the rear of the ambulance and held his hand in silence, but her eyes could not hide her concern and fear. The attending paramedic clipped onto the patient's left arm a small, modern device from which a flexible cable wire led to a digital display that was showing the irregular cardiac waveform. A warning sign in the upper right-hand corner of the display was flashing next to the low blood pressure reading. In a completely mechanical manner reflecting years of experience, the paramedic removed an adhesive patch from a plastic bag and attached it to Mr. Avon's right arm. The label on the discarded plastic package read "sterile microneedles." Then with her right hand, the paramedic inserted into the patch a narrow plastic tube while the fingers of her left hand proceeded to magically play the soft keys on the horizontal face of an electronic instrument. She dialed in an appropriate dosage of a new drug called Nocilis™. Within minutes, the display was showing a recovering cardiac waveform and the blood pressure warning faded into the dark green color of the screen. The paramedic looked with a smile at Mrs. Avon, who acknowledged her with a deep sigh of relief.

Lying in his hospital bed the next morning, Mr. Avon was slowly recovering from the disturbing events of the previous night. He knew that his youthful days were behind him, but the news from his physician that he needed a pacemaker could only cause him anguish. With an electronic stylus in his hand, he continued to record his thoughts and feelings on what appeared to be a synthetic white pad. The pen recognized the pattern of his handwriting and translated it to text for the laptop computer resting on the desk by the window. He drew a sketch of the pacemaker that Dr. Harte showed him in the morning; the computer stored an image of his lifesaving instrument. A little device barely the size of a silver dollar would forever remain in his chest and take control of his heart's rhythm. But a faint smile crossed Mr. Avon's lips when he remembered the doctor saying that the pacemaker would monitor his level of physical activity and correspondingly adjust his heart rate. He might be able to play tennis again, after all. With his remote control he turned on the projection screen television and slowly drifted back into light sleep.

This short fictional story illustrates how technology can touch our daily lives in so many different ways. The role of miniature devices and systems is not immediately apparent here because they are embedded deep within the applications they enable. The circumstances of this story called for such devices on many separate occasions. The miniature yaw-rate sensor in the vehicle stability system ensured that the ambulance would not skid on the icy highway. In the event of an accident, the crash acceleration sensor guaranteed that the airbags would deploy just in time to protect the passengers. The silicon manifold absolute pressure (MAP)

MEMS: A Technology from Lilliput

sensor in the engine compartment helped the engine's electronic control unit maintain, at the location's high altitude, the proper proportions in the mixture of air and fuel. As the vehicle was safely traveling, equally advanced technology in the rear of the ambulance saved Mr. Avon's life. The modern blood pressure sensor clipped onto his arm allowed the paramedic to monitor blood pressure and cardiac output. The microneedles in the adhesive patch ensured the immediate delivery of medication to the minute blood vessels under the skin, while a miniature electronic valve guaranteed the exact dosage. The next day, as the patient lay in his bed writing his thoughts in his diary, the microaccelerometer in the electronic quill recognized the motion of his hand and translated his handwriting into text. Another small accelerometer embedded in his pacemaker would enable him to play tennis again. He could also write and draw at will because the storage capacity of his disk drive was enormous, thanks to miniature read and write heads. And finally, as the patient went to sleep, an array of micromirrors projected a pleasant high-definition television image onto a suspended screen.

Many of the miniature devices listed in the above story, particularly the pressure and acceleration microsensors and the micromirror display, already exist as commercial products. Ongoing efforts at many companies and laboratories throughout the world promise to deliver, in the not-too-distant future, new and sophisticated miniature components and microsystems. It is not surprising, then, that there is widespread belief in the technology's future potential to penetrate far-reaching applications and markets.

What are MEMS—or MST?

In the United States, the technology is known as *microelectromechanical systems* (MEMS); in Europe it is called *microsystems technology* (MST). A question asking for a more specific definition is certain to generate a broad collection of replies, with few common characteristics other than "miniature." But such apparent divergence in the responses merely reflects the diversity of applications this technology enables, rather than a lack of commonality. MEMS is simultaneously a toolbox, a physical product, and a methodology all in one:

- It is a portfolio of techniques and processes to design and create miniature systems;

- It is a physical product often specialized and unique to a final application—one can seldom buy a generic MEMS product at the neighborhood electronics store;

- "MEMS is a way of making things," reports the Microsystems Technology Office of the United States Defense Advanced Research Program Agency (DARPA) [1]. These "things" merge the functions of sensing and actuation with computation and communication to locally control physical parameters at the microscale, yet cause effects at much grander scales.

Although a universal definition is lacking, MEMS products possess a number of distinctive features. They are miniature *embedded* systems involving one or many *micromachined* components or structures. They *enable* higher level functions, although in and of themselves their utility may be limited—a micromachined pressure sensor in one's hand is useless, but under the hood it controls the fuel-air mixture of the car engine. They often *integrate* smaller functions into one package for greater utility—for example, merging an acceleration sensor with electronic circuits for self-diagnostics. They can also bring *cost benefits,* directly through low unit pricing, or indirectly by cutting service and maintenance costs.

Although the vast majority of today's MEMS products are best categorized as components or subsystems, the emphasis in MEMS technology is on the "systems" aspect. True microsystems may still be a few years away, but their development and evolution rely on the success of today's components, especially as these components are integrated to perform functions ever increasing in complexity. Building microsystems is an evolutionary process. We spent the last thirty years learning how to build micromachined components. Only recently have we begun to learn about their seamless integration into subsystems, and ultimately into complete microsystems.

One notable example is the evolution of crash sensors for airbag safety systems. Early sensors were merely mechanical switches. They later evolved into micromechanical sensors that directly measured acceleration. The current generation of devices integrates electronic circuitry with a micromechanical sensor to provide self-diagnostics and a digital output. It is anticipated that the next generation of devices will also incorporate the entire airbag deployment circuitry that decides whether to inflate the airbag. As the technology matures, the airbag crash sensor may be integrated one day with micromachined yaw-rate and other inertial sensors to form a complete microsystem responsible for passenger safety and vehicle stability (Table 1.1).

Examples of future microsystems are not limited to automotive applications. Efforts to develop micromachined components for the control of fluids are just beginning to bear fruit. These could lead one day to the integration of micropumps with microvalves and reservoirs to build new miniature drug delivery systems.

Table 1.1
Examples of Present and Future Application Areas for MEMS

Commercial Applications	**Military Applications**
Invasive and noninvasive biomedical sensors	Inertial systems for munitions guidance and personal navigation
Miniature biochemical analytical instruments	Distributed unattended sensors for asset tracking, environmental and security surveillance
Cardiac management systems (e.g., pacemakers, catheters)	Weapons safing, arming, and fuzing
Drug delivery systems (e.g., insulin, analgesics)	Integrated micro-optomechanical components for identify-friend-or-foe systems
Neurological disorders (e.g., neurostimulation)	Head- and night-display systems
Engine and propulsion control	Low-power, high-density mass data storage devices
Automotive safety, braking, and suspension systems	Embedded sensors and actuators for condition-based maintenance
Telecommunication optical fiber components and switches	Integrated fluidic systems for miniature propellant and combustion control
Mass data storage systems	Miniature fluidic systems for early detection of biochemical warfare
Electromechanical signal processing	Electromechanical signal processing for small and low-power wireless communication
Distributed sensors for condition-based maintenance and monitoring structural health	Active, conformable surfaces for distributed aerodynamic control of aircraft
Distributed control of aerodynamic and hydrodynamic systems	

What is micromachining?

Micromachining is the set of design and fabrication tools that precisely machine and form structures and elements at a scale well below the limits of our human perceptive faculties—the microscale. Micromachining is the underlying foundation of MEMS fabrication; it is the toolbox of MEMS.

Arguably, the birth of the first micromachined components dates back many decades, but it was the well-established integrated circuit industry that indirectly played an indispensable role in fostering an environment suitable for the development and growth of micromachining technologies. As the following chapters will show, many tools used in the design and manufacturing of MEMS products are "borrowed" from the integrated circuit industry. It should not then be surprising that micromachining relies on silicon as a primary material, even though the technology was certainly demonstrated using other materials.

Applications and markets

Present markets are primarily in pressure and inertial sensors and inkjet print heads, with the latter dominated by Hewlett Packard Company of Palo Alto, California. Future and emerging applications include high-resolution displays, high-density data storage devices, valves, and fluid management and processing devices for chemical microanalysis, medical diagnostics, and drug delivery. While estimates for MEMS markets vary considerably, they all show significant present and future growth, reaching aggregate volumes in the many billions of dollars by the year 2004 [2–4]. The expected growth is driven by technical innovations and acceptance of the technology by an increasing number of end users and customers.

However, because of the lack of a single dominant application—the "killer app"—and the diverse technical requirements of end users, there is no single MEMS market, but rather a collection of markets, many of which are considered niche markets—especially when compared to their kin semiconductor businesses. It is true that unit volumes in a few segments, including automotive, are substantial, running in the tens of millions, but the corresponding dollar volumes tend to be modest. Furthermore, occasional poor forecasting of emerging applications poses additional risks and difficulties to companies engaged in the development and manufacture of MEMS products. For instance, the worldwide market for airbag crash sensors—thought by many to be a considerable market—is estimated today at $150,000,000, even as these components become standard on all 50,000,000 vehicles manufactured every year around the globe. Market studies conducted in the early 1990s incorrectly estimated the unit asking price of these sensors, neglecting the effect of competition on pricing, and artificially inflating the size of the market to $500,000,000 (Table 1.2).

To MEMS or not to MEMS?

Like many other emerging technologies with significant future potential, MEMS is subject to a rising level of excitement and publicity. As it evolves and end markets develop, this excessive optimism is gradually

Table 1.2
Analysis and Forecast of U.S. MEMS Markets (in Millions of U.S. Dollars)[1]

Year	Automotive[2] ($ 000,000)	Medical ($ 000,000)	Information Technology & Industrial[3] ($ 000,000)	Military & Aerospace ($ 000,000)	Total ($ 000,000)
1994	255.7	129.5	438.3	49.1	872.5
1995	298.0	146.1	459.0	54.8	957.9
1996	355.0	164.4	492.8	62.2	1,074.3
1997	419.0	187.0	527.0	71.6	1,204.6
1998	491.5	216.7	575.3	79.6	1,363.1
1999	562.0	245.7	645.9	95.8	1,549.4
2000	645.7	291.3	733.3	110.7	1,781.0
2001	758.5	354.8	836.0	133.3	2,082.5
2002	879.6	444.7	995.1	156.9	2,476.3
2003	1,019	562.9	1,222	176.7	2,980.4
2004	1,172	716.0	1,514	202.7	3,604.5
CAGR	16%	21%	16%	16%	17%

[1] Data prior to 1997 is actual. (The projected compound annual growth rate (CAGR) averages 17% across the dominant market sectors. Source: Frost & Sullivan [4].)
[2] Airbag systems and MAP sensors constitute 90% of the automotive MEMS market.
[3] In 1998, the market division was: inkjets 75.6%, displays 5.4%, and industrial 19%.

moderated with a degree of realism reflecting the technology's strengths and capabilities.

Any end user considering developing a MEMS solution or incorporating one into a design invariably reaches the difficult question of "Why MEMS?" The question strikes at the heart of the technology, particularly in view of competing methods, such as conventional machining or plastic molding techniques that do not have recourse to micromachining. For applications that can benefit from existing commercial MEMS products (e.g., pressure or acceleration sensors), the answer to the above question relies on the ability to meet required specifications and pricing. But the vast majority of applications require unique solutions that often necessitate the funding and completion of an evaluation or development program. In such situations, a clear-cut answer is seldom easy to establish.

In practice, a MEMS solution becomes attractive if it enables a new function, provides significant cost reduction, or both. For instance, medical applications generally seem to focus on added or enabled functionality and improved performance, whereas automotive applications often seek cost reduction. Size reduction can play an important selling role, but is seldom sufficient as the sole reason unless it becomes enabling itself. Naturally, reliability is always a dictated requirement. The decision-making process is further complicated by the fact that MEMS is not a single technology, but a set of technologies (e.g., surface vs. bulk micromachining). At this point, it is beneficial for the end user to become familiar with the capabilities and the limitations of any particular MEMS technology selected for the application in mind. The active participation of the end user allows for the application to drive the technology development, rather than the frequently occurring opposite situation.

Companies seeking MEMS solutions often contract a specialized facility for the design and manufacture of the product. Others choose first to evaluate basic conceptual designs through existing foundry services. A few companies may decide to internally develop a complete design. In the latter case, there is considerable risk that manufacturing considerations are not properly taken into account, resulting in significant challenges in production.

Regardless of how exciting and promising a technology may be, its ultimate realization is invariably dependent on economic success. The end user will justify the technology on the basis of added value, increased productivity, and/or cost competitiveness, and the manufacturer must

MEMS: A Technology from Lilliput

show revenues and profits. On both tracks, MEMS technology is able to deliver within a set of realistic expectations that may vary with the end application. A key element in cost competitiveness is "batch fabrication," which is the practice of simultaneously manufacturing hundreds or thousands of identical parts, thus diluting the overall impact of fixed costs, including the cost of maintaining expensive cleanroom and assembly facilities. This is precisely the same approach that has resulted, over the last few decades, in a dramatic decrease in the price of computer memory chips. Unfortunately, the argument works in reverse too: Small manufacturing volumes will bear the full burden of overhead expenses, regardless of how "enabling" the technology may be (Figure 1.1).

Standards

Few disagree that the burgeoning MEMS industry traces many of its roots to the integrated circuit industry. However, the two market dynamics differ greatly with severe implications, one of which is the lack of standards in MEMS. Complementary metal-oxide semiconductor (CMOS) technology has proven itself over the years to be a universally accepted manufacturing process for integrated circuits, driven primarily by the insatiable consumer demand for computers and digital electronics. In contrast, the lack of a dominant MEMS high-volume product, or family of products, combined with the unique technical requirements of

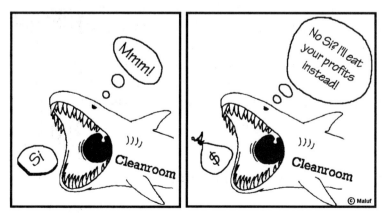

Figure 1.1 Volume manufacturing is essential for maintaining profitability.

each application have resulted in the emergence of multiple fabrication and assembly processes. The following chapters will introduce these processes. Standards are generally driven by the needs of high-volume applications, which are few in MEMS. In turn, the lack of standards feeds into the diverging demands of the emerging applications.

The psychological barrier

It is human nature to cautiously approach what is new because it is foreign and untested. Even for the technologically savvy or the fortunate individual living in high-tech regions, there is a need to overcome the comfort zone of the present before engaging the technologies of the future. This cautious behavior translates to slow acceptance of new technologies and derivative products as they are introduced into society. MEMS acceptance is no exception. For example, demonstration of the first micromachined accelerometer took place in 1979 at Stanford University. Yet it took nearly fifteen years before it became accepted as a device of choice for automotive airbag safety systems. Naturally, in the process, it was designed and redesigned, tested and qualified in the laboratory and the field before it gained the confidence of automotive suppliers. The process can be lengthy, especially for embedded systems (see Figure 1.2).

Today, MEMS and associated product concepts generate plenty of excitement, but not without skepticism. Companies exploring for the first time the incorporation of MEMS solutions into their systems do so with trepidation, until an internal "MEMS technology champion" emerges to educate the company and raise the confidence level. With many micromachined silicon sensors embedded in every car and in numerous critical medical instruments, and with additional MEMS products finding their way into our daily lives, the height of this hidden psychological barrier appears to be declining.

Journals, conferences, and Web sites

The list of journals and conferences focusing on micromachining and MEMS continues to grow every year. There is also a growing list of on-

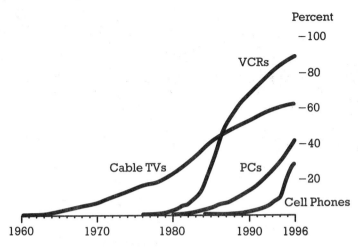

Figure 1.2 Chart illustrating the percent household penetration of new electronic products. It takes 5 to 15 years before new technologies gain wide acceptance [5].

line Web sites, most notably the MEMS Clearinghouse hosted by the Information Sciences Institute (ISI), Marina del Rey, California, and the European Microsystems Technology On-line (EMSTO), Berlin, Germany, sponsored by the ESPRIT program of the European Commission. The sites provide convenient links and maintain relevant information directories (Table 1.3).

List of journals and magazines

Several journals and trade magazines published in the U.S. and Europe cover research and advances in the field. Some examples are:

- *Sensors and Actuators* (A, B & C): a peer-reviewed scientific journal published by Elsevier Science, Amsterdam, The Netherlands.
- *Journal of Micromechanical Systems* (JMEMS): a peer-reviewed scientific journal published by the Institute of Electrical and Electronic Engineers (IEEE), Piscataway, New Jersey, in collaboration with the American Society of Mechanical Engineers (ASME), New York, New York.
- *Journal of Micromechanics and Microengineering* (JMM): a peer-reviewed scientific journal published by the Institute of Physics, Bristol, United Kingdom.

Table 1.3
List of a Few Government and Nongovernment
Organizations With Useful On-line Resources

Organization	Address	Description	Web Site
MEMS ISI Clearinghouse	Marina del Rey, CA	U.S. Clearinghouse	mems.isi.edu
EMSTO	Berlin, Germany	European Clearinghouse	www.nexus-emsto.com
VDI/VDE – IT	Teltow, Germany	Association of German Engineers	www.vdivde-it.de/MST
DARPA	Arlington, VA	Sponsored U.S. government projects	web-ext2.darpa.mil/MTO
NIST	Gaithersburg, MD	Sponsored U.S. government projects	www.atp.nist.gov
Institute of Defense Analyses	Alexandria, VA	Insertion in military applications	www.ida.org/MEMS
AIST – MITI	Tokyo, Japan	The "Micromachine Project" in Japan	www.aist.go.jp

- *Sensors Magazine*: a trade journal with an emphasis on practical and commercial applications published by Helmers Publishing Inc., Peterborough, New Hampshire.

- *MST news:* an international newsletter on microsystems and MEMS published by VDI/VDE Technologiezentrum Informationstechnik GmbH, Teltow, Germany.

- *Micromachine Devices*: a publication companion to *R&D Magazine* with news and updates on MEMS technology published by Cahners Business Information, Des Plaines, Illinois.

List of conferences and meetings

Several conferences cover advances in MEMS or incorporate program sessions on micromachined sensors and actuators. The following list gives a few examples:

- International Conference on Solid-State Sensors and Actuators (Transducers): held on odd years and rotates sequentially between North America, Asia, and Europe.

- Solid-State Sensor and Actuator Workshop (Hilton Head): held on even years in Hilton Head Island, South Carolina, and sponsored by the Transducers Research Foundation, Cleveland, Ohio.
- MicroElectroMechanical Systems Workshop (MEMS): an international meeting held annually and sponsored by the Institute of Electrical and Electronics Engineers (IEEE), Piscataway, New Jersey.
- International Society for Optical Engineering (SPIE): regular conferences held in the United States and sponsored by SPIE, Bellingham, Washington.
- MicroTotalAnalysis Systems (MTAS): a conference focusing on microanalytical and chemical systems. This conference was held on alternating years, but will become annual beginning in the year 2000. It alternates between North America and Europe.

Summary

Microelectromechanical structures and systems are miniature devices that enable the operation of complex systems. They exist today in many environments, especially automotive, medical, consumer, industrial, and aerospace. Their potential for future penetration into a broad range of applications is real, supported by strong developmental activities at many companies and institutions. The technology consists of a large portfolio of design and fabrication processes (a toolbox), many borrowed from the integrated circuit industry. The development of MEMS is inherently interdisciplinary, necessitating an understanding of the toolbox as well as the end application.

References

[1] Dr. Albert Pisano, in presentation material distributed by the United States Defense Advanced Research Program Agency (DARPA), available at http://web-ext2.darpa.mil.

[2] System Planning Corporation, "Microelectromechanical Systems (MEMS): An SPC Market Study," January 1999, 1429 North Quincy Street, Arlington, Virginia 22207.

[3] Frost & Sullivan, "World Sensors Market: Strategic Analysis," Report #5509–32, February 1999, 2525 Charleston Road, Mountain View, California 94043, http://www.frost.com.

[4] Frost & Sullivan, "U.S. Microelectromechanical Systems (MEMS)," Report #5549–16, June 1997, 2525 Charleston Road, Mountain View, California 94043, http://www.frost.com.

[5] Mercer Management Consulting, Inc., In *Business Week*, April 19, 1999, p. 8.

Selected bibliography

Angell, J. B., Terry, S. C., and Barth, P. W., "Silicon Micromechanical Devices," *Scientific American*, Vol. 248, No. 4, Apr. 1983, pp. 44–55.

Gabriel, K. J., "Engineering Microscopic Machines," *Scientific American*, Vol. 273, No. 3, Sept. 1995, pp. 150–153.

Petersen, K. E., "Silicon as a Mechanical Material," *Proceedings of the IEEE*, Vol. 70, No. 5, May 1982, pp. 420–457.

"Nothing but light," *Scientific American*, Vol. 279, No. 6, Dec. 1998, pp. 17–20.

CHAPTER 2

Contents

Silicon material system

Other materials and substrates

Important material properties and physical effects

Summary

The Sandbox: Materials for MEMS

You can't see it, but it's everywhere you go.

Bridget Booher, journalist, on silicon.

If we view micromachining technology as a set of generic tools, then there is no reason to limit its use to one material. Indeed, micromachining was demonstrated in silicon, glass, ceramics, polymers, and compound semiconductors made of group III and V elements, as well as a wide variety of metals including titanium and tungsten. However, silicon remains the primary material of choice for microelectromechanical systems. Unquestionably, this popularity arises from the large momentum of the electronic integrated circuit industry and the derived economic benefits, not the least of which is the extensive industrial infrastructure. The object of this chapter is to present the properties of silicon and a few other materials while emphasizing that the final choice of materials is determined by the type of application and economics.

Silicon material system

The silicon material system encompasses, in addition to silicon itself, a host of materials commonly used in the semiconductor integrated circuit industry. Normally deposited as thin films, they include silicon oxides, nitrides, and carbides as well as metals such as aluminum, titanium, tungsten, and copper.

Silicon

Silicon is one of very few materials that can be economically manufactured in single crystal substrates. This crystalline nature provides significant electrical and mechanical advantages. The precise modulation of silicon's electrical conductivity using impurity doping lies at the very core of the operation of electronic semiconductor devices. Mechanically, silicon is an elastic and robust material whose characteristics have been very well studied and documented. The tremendous wealth of information accumulated on silicon and its compounds over the last few decades has made it possible to innovate and explore new areas of application extending beyond the manufacturing of electronic integrated circuits. It becomes evident that silicon is a suitable material platform on which electronic, mechanical, thermal, optical, and even fluid flow functions can be integrated (see Table 2.1). Ultrapure electronic-grade silicon wafers available for the integrated circuit industry are common today in MEMS. The low cost of these substrates (approximately $10 for a 100-mm diameter wafer) makes them attractive for the fabrication of micromechanical components and systems.

Silicon as an element exists in any of three forms: *Crystalline, polycrystalline*, or *amorphous*. Polycrystalline, or simply polysilicon, and amorphous silicon are usually deposited as thin films with typical thicknesses below 5 μm. Crystalline silicon substrates are commercially available as circular wafers with either 100-mm (4 in.) or 150-mm (6 in.) diameters. Larger diameter (200-mm and 300-mm) wafers are currently economically unjustified for MEMS fabrication and their use is strictly for the integrated circuit industry. Standard 100-mm wafers are nominally 525-μm-thick and 150-mm wafers are typically 650-μm-thick. Double sided, polished wafers commonly used for micromachining on both sides of the wafer, are approximately 100 μm thinner than standard thickness substrates.

The Sandbox: Materials for MEMS

Table 2.1
Properties of a Selected List of Materials

Property	Si	SiO$_2$	Si$_3$N$_4$	Quartz	SiC	AlN	92% Al$_2$O$_3$
Relative permittivity (ε_0)	11.8	3.8	4	3.75	9.7	8.5	9
Dielectric strength (V/cm $\times 10^6$)	3	5–10	5–10	25–40	4	13	11.6
Electron mobility (cm^2/V · s)	1500	—	—	—	1000	—	—
Hole mobility (cm^2/V · s)	400	—	—	—	40	—	—
Young's modulus (GPa)	160	73	323	107	450	340	275
Yield strength (GPa)	7	8.4	14	9	21	16	15.4
Poisson's ratio	0.22	0.17	0.25	0.16	0.14	0.31	0.31
Density (g/cm^3)	2.4	2.3	3.1	2.65	3.2	3.26	3.62
Coefficient of thermal expansion (10^{-6}/° C)	2.6	0.55	2.8	0.55	4.2	4.0	6.57
Thermal conductivity at 300K (W/cm · K)	1.57	0.014	0.19	0.0138	5	1.60	0.36
Specific heat (J/g · K)	0.7	1.0	0.7	0.787	0.8	0.71	0.8
Melting temperature (° C)	1415	1700	1800	1610	2830	2470	1800

Proper visualization of crystallographic planes is key to understanding the dependence of material properties on crystal orientation and the effects of plane-selective etch solutions. Silicon has a diamond lattice crystal structure that can be regarded as simple cubic. In other words, the primitive unit—the smallest repeating block—of the crystal lattice resembles a cube. The three major coordinate axes of the cube are called the "principal axes." Specific directions and planes within the crystal are designated in reference to the principal axes using "Miller indexes"[1], a special notation that includes three integers enclosed in carets, brackets, parentheses, and braces. Directions are specified by brackets or carets, whereas planes are defined with parentheses or braces (Figure 2.1). For example, [100] represents a specific vector direction (the +x-direction) referred to the three principal axes (x,y,z) of the cube, and <100> denotes the six directions equivalent to [100] (the +x, -x, +y, -y, +z, and -z directions). Similarly, (111) is a plane perpendicular to the [111] vector (a diagonal vector through two farthest corners), whereas {111} represents all eight equivalent (111) crystallographic planes. What determine plane and direction equivalence are the symmetry operations that carry a crystal lattice (including the primitive unit) back into itself, i.e., the transformed lattice after the symmetry operation is complete is identical to the starting lattice. With some thought, it becomes clear that 90 degree rotations and mirror operations about the three principal axes are symmetry operations for a simple cubic crystal. Therefore, the +x direction is equivalent to the +y direction under a 90° rotation; the +y direction is equivalent to the −y direction under a mirror operation, and so forth. Hence, the +x, −x, +y, −y, +z, and −z directions are all equivalent. Vector algebra can show that the angles between {100} and {110} planes, and between {100} and {111} planes are 45° and 54.74°, respectively. Similarly, {111} and {110} planes can intersect each other at 35.26°, 90°, or 144.74°. The angle between {100} and {111} planes is of particular importance in micromachining because many alkaline aqueous solutions, such as potassium hydroxide (KOH), selectively etch the {100} planes of silicon but not the {111} planes. The etch results in cavities that are bounded by {111} planes (Figure 2.1).

Material manufacturers cut thin circular wafers from large silicon boules along specific crystal planes. The cut plane—the top surface of the wafer—is known as the orientation cut, and is encoded on the wafer itself in the form of a primary and a secondary flat located on the sides of the wafer. For instance, the top surface of a {100}-orientation wafer is a {100} plane, which could be any of the six equivalent (100) planes. Commercially available wafers are predominantly of {100} orientation, which is the preferred orientation cut for CMOS technology. In addition to orientation cut, impurity doping type (n or p) and electrical resistivity (in $\Omega \cdot cm$) are also specified by the supplier (Figure 2.2).

Crystalline silicon is a hard and brittle material deforming elastically until it reaches its yield strength, at which point it breaks. Its tensile yield strength is 7 GPa equivalent to a 700-kg (1500 lb.) weight suspended from a 1 mm^2 area. Its Young's modulus is dependent on crystal orientation with an average value of 160 GPa, near that of stainless steel. The dependence of the mechanical properties on crystalline orientation is reflected in the way a silicon wafer preferentially cleaves along crystal planes. While large silicon wafers tend to be fragile, individual dice with dimensions on the order of 1 cm × 1 cm or less are rugged and can sustain relatively harsh handling conditions. As a direct consequence of

The Sandbox: Materials for MEMS

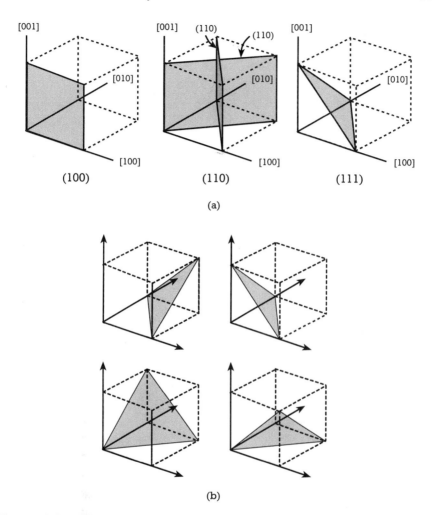

Figure 2.1 (a) Three crystallographic planes and their Miller indexes for a simple cubic crystal. Two equivalent (110) planes in the {110} set of planes are identified. (b) Four of the eight equivalent (111) planes in the {111} family.

crystalline nature, mechanical properties are uniform across wafer lots, and wafers are free of intrinsic stresses. This helps to minimize the number of design iterations for silicon transducers that rely on stable mechanical properties for their operation. Bulk mechanical properties of crystalline silicon are largely independent of impurity doping, but stresses tend to rise when dopant concentrations reach high levels ($\sim 10^{20}$ cm^{-3}).

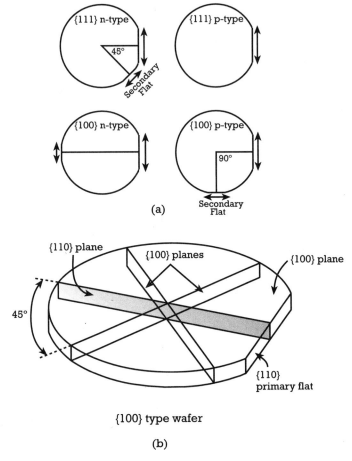

Figure 2.2 (a) Illustration showing the primary and secondary flats of {100} and {111} wafers for both n-type and p-type doping (SEMI standard). (b) Illustration identifying various planes in a wafer of {100} orientation.

Polysilicon is an important material in the integrated circuit industry and has been extensively studied. A detailed description of its electrical properties is found in Kamins [2]. Polysilicon is an equally important and attractive material for MEMS. It has been successfully used to make micromechanical structures and to integrate electrical interconnects, thermocouples, p-n junction diodes, and many other electrical devices with micromechanical structures. The most notable example is the acceleration sensor for airbag safety systems, available from Analog Devices Inc., Norwood, Massachusetts. Surface micromachining based on

The Sandbox: Materials for MEMS

polysilicon is today a well-established technology for forming thin (a few micrometers thick), and planar devices.

The mechanical properties of polycrystalline and amorphous silicon vary with deposition conditions, but, for the most part, they are similar to those of single-crystal silicon [3]. Both normally suffer from relatively high levels of intrinsic stress (hundreds of MPa) which requires annealing at elevated temperatures (> 900° C). Beam structures made of polycrystalline or amorphous silicon that have not been subjected to a careful stress-annealing step can curl under the effect of intrinsic stress (Table 2.2).

Silicon is a good thermal conductor with a conductivity approximately one hundred times larger than that of glass. In complex integrated systems, the silicon substrate can be used as an efficient heat sink. This feature will be revisited when we review thermal-based sensors and actuators.

Unfortunately, silicon is not an active optical material—silicon-based lasers do not exist. Because of the particular interactions between the crystal atoms and the conduction electrons, silicon is effective only in detecting light; emission of light is very difficult to achieve. At infrared wavelengths above 1.1 μm silicon is transparent, but at wavelengths shorter than 0.4 μm (in the blue and ultraviolet portions of the visible spectrum), it reflects over 60% of the incident light (Figure 2.3). The

Table 2.2
Temperature Dependence of Some Material Properties of Crystalline Silicon [4].

	300 K	400 K	500 K	600 K	700 K
Coefficient of linear expansion (10^{-6} K^{-1})	2.616	3.253	3.614	3.842	4.016
Specific heat (J/g · K)	0.713	0.785	0.832	0.849	0.866
Thermal conductivity (W/cm · K)	1.56	1.05	0.8	0.64	0.52
Temperature coefficient of Young's Modulus (10^{-6} K^{-1})	−90	−90	−90	−90	−90
Temperature coefficient of piezoresistance (10^{-6} K^{-1}) (doping < 10^{18} cm^{-3})	−2500	−2500	−2500	—	—
Temperature coefficient of permittivity (10^{-6} K^{-1})	1000	—	—	—	—

attenuation depth of light in silicon (the distance light travels before the intensity drops to 36% of its initial value) is 2.7 μm at 633 nm (red), and 0.2 μm at 436 nm (blue). The slight attenuation of red light relative to other colors is what gives thin silicon membranes their translucent reddish tint (Figure 2.3).

Silicon is also known to retain its mechanical integrity at temperatures up to about 500° C. At higher temperatures, silicon softens appreciably and plastic deformation sets in. While the mechanical and thermal properties of polysilicon are similar to those of single crystal silicon, polysilicon experiences slow-stress-annealing effects at temperatures above 250° C, making its operation at elevated temperatures subject to long-term instabilities, drift, and hysteresis effects.

The interactions of silicon with gases, chemicals, biological fluids, and enzymes remain the subject of many research studies, but for the most part, silicon is considered stable and resistant to many elements and chemicals typical of daily applications. For example, experiments have shown that silicon remains intact in the presence of Freon™ gases as well as corrosive automotive fluids such as brake fluids. Silicon has also proven to be a suitable material for applications involving the delivery of ultra-high-purity gases. In medicine and biology, studies are ongoing to evaluate silicon for chronic medical implants. Preliminary medical evidence indicates that silicon is benign in the body and does not release toxic substances when it comes in contact with biological fluids.

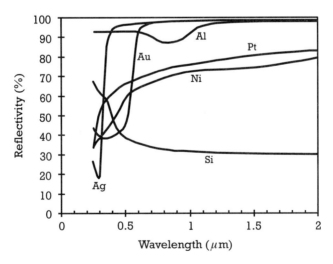

Figure 2.3 Optical reflectivity for silicon and a selected list of metals.

The Sandbox: Materials for MEMS

However, it appears from recent experiments that bare silicon surfaces may not be suitable for high-performance polymerase chain reactions (PCR) intended for the amplification of genetic DNA material.

Silicon oxide and nitride

It is often argued that silicon is such a successful material because it has a stable oxide that is electrically insulating, unlike germanium whose oxide is soluble in water, or gallium arsenide whose oxide cannot be grown appreciably. Various forms of silicon oxides (SiO_2, SiO_x, silicate glass) are widely used in micromachining, due to their excellent electrical and thermal insulating properties. They are also used as sacrificial layers in surface micromachining processes because they can be preferentially etched in hydrofluoric acid (HF) with high selectivity to silicon. Silicon dioxide (SiO_2) is thermally grown by oxidizing silicon at temperatures above 800° C, whereas the other forms of oxides and glass are deposited by chemical vapor deposition, sputtering, or even spin-on (the various deposition ways will be described in the next chapter). Silicon oxides and glass layers are known to soften and flow when subjected to temperatures above 700° C. A drawback of silicon oxides is their relatively large intrinsic stresses that are difficult to control or anneal. This has limited their use as materials for large suspended beams or membranes.

Silicon nitride (Si_xN_y) is also a widely used insulating thin film and is effective as a barrier against mobile ion diffusion, in particular, sodium and potassium ions found in biological environments. Its Young's modulus is higher than that of silicon and its intrinsic stresses can be controlled by the specifics of the deposition process. Silicon nitride is an effective masking material in many alkaline etch solutions.

Thin metal films

The choice of a thin metal film depends greatly on the nature of the final application (Table 2.3). Thin metal films are normally deposited either by sputtering, evaporation, or chemical vapor deposition; gold, nickel and Permalloy™ (Ni_xFe_y) can also be electroplated (Table 2.3).

For basic electrical interconnections, aluminum is the most common and is relatively easy to deposit by sputtering, but its operation is limited to noncorrosive environments and to temperatures below 300° C. For higher temperatures and harsher environments, gold, titanium, and

Table 2.3
List of Selected Metals That Can Be Deposited as Thin Films
(Up to a Few μm in Thickness) with Corresponding Electrical
Resistivities and Typical Areas of Application

Metal	ρ ($\mu\Omega \cdot$ cm)	Typical Areas of Application
Ag	1.58	Electrochemistry
Al	2.7	Electrical interconnects
		Optical reflection in the visible and the infrared
Au	2.4	High temperature electrical interconnects
		Optical reflection in the infrared
		Electrochemistry
Cr	12.9	Intermediate adhesion layer
Cu	1.7	Low resistivity electrical interconnects
Indium-tin oxide (ITO)	300–3,000	Transparent conductive layer for liquid crystal displays
Ir	5.1	Electrochemistry
		Microelectrodes for sensing biopotentials
Ni	6.8	Magnetic transducing
NiCr	200–500	Thin film laser-trimmed resistor
Pd	10.8	Electrochemistry
		Solder wetting layer
Permalloy™ (Ni_xFe_y)	—	Magnetic transducing
Pt	10.6	Electrochemistry
		Microelectrodes for sensing biopotentials
SiCr	2,000	Thin film laser-trimmed resistor
SnO_2	5,000	Chemoresistance in gas sensors
TaN	300–500	Negative temperature coefficient of resistance (TCR)
		Thin film laser-trimmed resistor
Ti	42	Intermediate adhesion layer
TiNi	80	Shape-memory alloy, actuation
TiW	75–200	Intermediate adhesion layer
		Near zero temperature coefficient of resistance (TCR)
W	5.5	High temperature electrical interconnects

tungsten are excellent substitutes. Aluminum tends to anneal over time with temperature causing changes in its intrinsic stresses. As a result, it is typically located away from stress- or strain-sensing elements. Aluminum is a good light reflector in the visible, and gold excels in the infrared.

The Sandbox: Materials for MEMS

Platinum and palladium are two very stable materials for electrochemistry, though their fabrication entails some added complexity. Gold, platinum, and iridium are good choices for microelectrodes used in electrochemistry and in sensing biopotentials. Silver is also useful in electrochemistry. Chromium, titanium, and titanium-tungsten are frequently used as very thin (10–100 nm) adhesion layers for highly stressed metals with a tendency to peel off, such as sputtered or evaporated tungsten, nickel, platinum, or palladium. Metal bi-layers consisting of an adhesion layer (e.g., chromium) and an intermediate nickel or platinum layer are normally used to solder with silver-tin or tin-lead alloys. For applications requiring transparent electrodes, such as liquid crystal displays, indium-tin-oxide (ITO) meets the requirements. Finally, Permalloy™ has been explored as a material for thin magnetic cores.

Polymers

Polymers, in the form of polyimides or photoresist, can be deposited with varying thicknesses from a few nanometers to hundreds of microns. Standard photoresist is spin-coated to a thickness of 1 to 10 μm, but special photoresists such as the epoxy-based SU-8 [5] can form layers up to 100-μm-thick. Hardening of the resist under ultraviolet light produces rigid structures. Spin-on organic polymers are generally limited in their application because they shrink substantially after the solvent evaporates, and because they cannot sustain temperatures above 200° C. Because of their unique absorption and adsorption properties, polymers have gained acceptance in the sensing of chemical gases and humidity [6].

Other materials and substrates

Over the years, micromachining methods were applied to a variety of substrates to fabricate passive microstructures and transducers. Fabrication processes for glass and quartz are mature and well established but for other materials, such as silicon carbide, new techniques are being explored and developed. In the process, these activities add breadth to micromachining technology and enrich the inventory of available tools. The following sections briefly review the use of a few materials other than silicon.

Glass and quartz substrates

Glass is, without a doubt, a companion material to silicon. Both are bonded together figuratively and literally in many ways. Silicon originates from processed and purified silicates, a form of glass, and it can be made to bond electrostatically to Pyrex® glass substrates—a process called anodic bonding that is common in the making of pressure sensors. But, like all relatives, differences remain. Glass generally has a higher coefficient of thermal expansion resulting in interfacial stresses between bonded silicon and glass substrates.

Micromachining of glass and quartz substrates is practical in special applications, such as when an optically transparent or an electrically insulating substrate is required. Quartz also has the distinct property of being piezoelectrical. However, micromachining of glass or quartz is limited in scope relative to silicon. Etching in hydrofluoric acid (HF) or ultrasonic drilling typically yields coarsely defined features with poor edge control. Thin metal films can be readily deposited on glass or quartz substrates and defined, using standard lithographic techniques. Channels microfabricated in glass substrates with thin metal microelectrodes have been useful in making capillaries for miniaturized biochemical analysis systems.

Silicon carbide and diamond

Silicon carbide and diamond continue to captivate the imagination of many in the micromachining community. Both materials offer significant advantages, particularly hardness, resistance to harsh environments, and excellent thermal properties. Some micromachining in silicon carbide [7] and diamond has been demonstrated, however, much remains to be studied about both materials and their potential use in MEMS. An important feature of both silicon carbide and diamond is that they exhibit piezoresistive properties. High temperature pressure sensors in silicon carbide substrates were developed, with stable operation up to about 500° C, but it is likely that commercialization remains many years away.

Silicon carbide substrates are available in polycrystalline form, but they are typically expensive and are available in rather small diameters. Instead, silicon carbide is grown or deposited on silicon substrates. Crystalline silicon carbide can be obtained by epitaxial growth directly on silicon, but the material is generally of poor quality suffering from voids and dislocations. Polycrystalline thin films deposited by chemical vapor deposition have received some interest for applications involving harsh environments, especially as a coating material.

Diamond is an even lesser explored material than silicon carbide. Thin synthetic diamond films made with thicknesses up to a few microns can be achieved using chemical vapor deposition, but their use remains limited to academic and research laboratories.

Gallium arsenide and other group III-V compound semiconductors

Instead of pondering the utility of gallium arsenide (GaAs) and other group III-V compounds (e.g., InP, AlGaAs, GaN) as alternate substrate materials to silicon, it is perhaps more appropriate to think of micromachining as a set of tools that can provide solutions to issues specific to devices that currently can only be built in these materials, particularly lasers and optical devices. In that regard, micromachining becomes an application-specific toolbox whose main characteristic is to address ways that can enable new functions or enhance existing ones. Basic micromechanical structures such as beams were formed in gallium arsenide, but micromachining is proving to be more useful at devising ways to incorporate micromirrors in resonant optical cavities for tunable lasers. Moreover, micromachining using gallium arsenide and group III-V compound semiconductors is a practical way to integrate RF switches, antennas, and other custom high-frequency components with ultra-high-speed electronic devices for wireless telecommunications.

Shape-memory alloys

The shape-memory effect is a unique property of a special class of alloys that return to a predetermined shape when heated above a critical "transition temperature." The material "remembers" its original shape after being strained and deformed. The discovery was first made in a gold-cadmium alloy in 1951, but was quickly extended to a broad range of other alloys including titanium-nickel, copper-aluminum-nickel, iron-nickel, and iron-platinum alloys. A basic understanding of the underlying physical principles was established in the 1970s, but extensive research remains ongoing in an effort to develop a thorough theoretical foundation. Nonetheless, the potential applications for shape-memory alloys abound. It has been estimated that upwards of 15,000 patents have been applied for on this topic. Titanium-nickel alloys have been the most widely used of shape-memory alloys because of their relatively simple composition and robustness.

An important factor that determines the practical utility of the alloy is its transition temperature. Below this temperature, it has a low yield strength; in other words, it is readily deformed into new permanent shapes. The deformation can be 20 times larger than the elastic deformation with no permanent strain. When heated above its transition temperature, the material completely recovers its original shape through complex changes in its crystal structure. The process generates very large forces making shape-memory alloys ideal for actuation purposes. In contrast, piezoelectric and electrostatic actuators exert only a fraction of the force available from a shape-memory alloy, but they act much more quickly.

Bulk titanium-nickel alloys in the form of wires and rods are commercially available under the name Nitinol™ [8]. Its transition temperature can be tailored between -100 and $100°$ C, typically by controlling impurity concentration. Recently, thin titanium-nickel films with thicknesses up to $50\,\mu$m were successfully demonstrated with properties similar to those of Nitinol™. Titanium-nickel is a good electrical conductor with a resistivity of $80\,\mu\Omega \cdot$ cm, but a relatively poor thermal conductor with a conductivity about one-tenth that of silicon. Its yield strength is only 100 MPa below its transition temperature, but rapidly increases to 560 MPa when heated above it. The Young's modulus shows a similar dependence on temperature; at low temperatures it is 28 GPa, increasing to 75 GPa above the transition temperature.

Important material properties and physical effects

The interaction of physical parameters with each other, most notably electricity with mechanical stress, temperature and thermal gradients, magnetic fields, and incident light, yields a multitude of phenomena of great interest to MEMS. We will briefly review in this section three commonly used effects: Piezoresistivity, piezoelectricity, and thermoelectricity.

The Sandbox: Materials for MEMS

Piezoresistivity

Piezoresistivity is a widely used physical effect that has its name derived from the Greek word *piezein,* meaning to apply pressure. Discovered first by Lord Kelvin in 1856, it is the phenomenon by which an electrical resistance changes in response to mechanical stress. The first application of the piezoresistive effect was metal strain gauges to measure strain from which other parameters such as force, weight, and pressure were inferred (Figure 2.4). C. S. Smith's discovery in 1954 [9] that silicon and germanium had a much greater piezoresistive effect than metals spurred significant interest. The first pressure sensors based on diffused (impurity-doped) resistors in thin silicon diaphragms were demonstrated in 1969 [10]. The vast majority of today's commercially available pressure sensors use silicon piezoresistors.

For the physicist at heart, piezoresistivity arises from the deformation of the energy bands as a result of an applied stress. In turn, the deformed bands affect the effective mass and the mobility of electrons and holes, hence modifying resistivity. For the engineer at heart, the fractional change in resistivity, $\Delta\rho/\rho$, is to a first order linearly dependent on $\sigma_{//}$ and $\sigma_\perp$, the two stress components parallel and orthogonal to the direction of the resistor, respectively. The direction of the resistor is here defined as that of the current flow. The relationship can be expressed as:

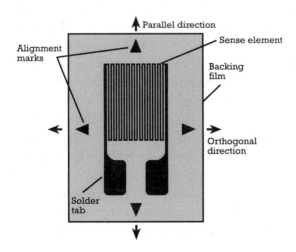

Figure 2.4 A typical thin metal foil strain gauge mounted on a backing film. Stretching of the sense element causes a change in its resistance.

$$\Delta\rho/\rho = \pi_{//}\sigma_{//} + \pi_{\perp}\sigma_{\perp}$$

where the proportionality constants, $\pi_{//}$ and $\pi_{\perp}$, are called the parallel and perpendicular piezoresistive coefficients, respectively, and are related to the gauge factor[1] by the Young's modulus of the material. The piezoresistive coefficients are dependent on crystal orientation and can change significantly from one direction to the other. They also depend on dopant type (n-type vs. p-type) and concentration. For {100} wafers, the piezoresistive coefficients for p-type elements are maximal in the <110> directions and vanish along the <100> directions. In other words, p-type piezoresistors must be oriented along the <110> directions to measure stress, and thus should be either aligned or perpendicular to the wafer primary flat. Those at 45° with respect to the primary flat, i.e., in the <100> direction, are insensitive to applied stress, which provides an inexpensive way to incorporate stress-independent diffused temperature sensors. The crystal orientation-dependence of the piezoresistive coefficients takes a more complex function for piezoresistors diffused in {110} wafers, but this dependence fortuitously disappears in {111} wafers. More descriptive details of the underlying physics of piezoresistivity and dependence on crystal orientation can be found in Kanda [11], and Middelhoek and Audet [12] (Table 2.4).

If we consider p-type piezoresistors diffused in {100} wafers and oriented in the <110> direction (parallel or perpendicular to the flat), it is apparent from the positive sign of $\pi_{//}$ in Table 2.4 that the resistance increases with tensile stress applied in the parallel direction, $\sigma_{//}$, as if the piezoresistor itself is being elongated. Furthermore, the negative sign of $\pi_{\perp}$ implies a decrease in resistance with tensile stress orthogonal to the resistor, as if its width is being stretched. In actuality, the stretching or contraction of the resistor is not the cause of the piezoresistive effect, but they make a fortuitous analogy to readily visualize the effect of stress on resistance. This analogy breaks down for n-type piezoresistors.

Like many other physical effects, piezoresistivity is a strong function of temperature. For lightly doped silicon (n- or p-type, $< 10^{18}$ cm^{-3}), the temperature coefficient of $\pi_{//}$ and $\pi_{\perp}$ is approximately 0.25% per °C. It decreases with dopant concentration to ~ 0.1% per °C at 8×10^{19} cm^{-3}.

---

1. The gauge factor, K, is the constant of proportionality relating the fractional change in resistance, $\Delta R/R$, to the applied strain, ε, by the relationship $\Delta R / R = K \cdot \varepsilon$.

The Sandbox: Materials for MEMS

Table 2.4
Piezoresistive Coefficients for n- and p-type {100} Wafers and Doping Levels Below 10^{18} cm^{-3}.*

	$\pi_\parallel$ (10^{-13} m^2/N)	$\pi_\perp$ (10^{-13} m^2/N)	
p-type	0	0	in <100> direction
	72	−65	in <110> direction
n-type	−102	53	in <100> direction
	−32	0	in <110> direction

* The values decrease precipitously at higher doping concentrations

Polysilicon and amorphous silicon also exhibit a strong piezoresistive effect. A wide variety of sensors using polysilicon piezoresistive sense elements have been demonstrated. Clearly, piezoresistive coefficients lose their sensitivity to crystalline direction and become an average over all orientations. Instead, the gauge factor, K, relating the fractional change in resistance to strain is often used. Gauge factors in polysilicon and amorphous silicon range typically between 20 and 40, about a factor of five smaller than in single crystalline silicon. The gauge factor decreases quickly as doping concentration exceeds 10^{19} cm^{-3}. However, one advantage of polysilicon over crystalline silicon is its reduced temperature coefficient of resistance (TCR). At doping levels approaching 10^{20} cm^{-3}, the TCR for polycrystalline silicon is approximately 0.04% per °C compared to 0.14% per °C for crystalline silicon. The deposition process and the dopant species have been found to even alter the sign of the TCR. For example, emitter-type polysilicon (a special process for depositing heavily doped polysilicon to be used as an emitter for bipolar transistors) has a TCR of −0.045% per °C. Resistors with negative TCR are particularly useful in compensating the positive temperature dependence of piezoresistive sensors.

Piezoelectricity

Certain classes of crystals exhibit the peculiar property of producing an electric field when subjected to an external force. They also expand or contract in response to an externally applied voltage. The effect was discovered in quartz by the brothers Pierre and Jacques Curie in 1880 [13]. Its first practical application was in the 1920s when a Frenchman,

Langevin, developed a quartz transmitter and receiver for underwater sound—the first Sonar! Piezoelectric crystals are common in many modern applications, for example, as clock oscillators in computers and as ringers in cellular telephones. They are attractive for MEMS because they can be used as sensors as well as actuators, and they can be deposited as thin films over standard silicon substrates.

The physical origin of piezoelectricity is explained by charge asymmetry within the primitive unit resulting in the formation of a net electric dipole. Adding up these individual dipoles over the entire crystal gives a net polarization and an effective electric field within the material. Crystal symmetry again plays an important role: only a crystal that lacks a center of symmetry exhibits piezoelectric properties. A crystal with a center of symmetry, such as a cubic crystal, is not piezoelectric because the net electric dipole within the primitive unit is always vanishing, even in the presence of an externally applied stress (Figure 2.5). Silicon is not piezoelectric because it is cubic and, further, it is held together by covalent (not ionic) bonding.

If we consider an ionic or partly ionic crystal lacking a center of symmetry, for example zinc oxide (ZnO), the net electric dipole internal to the primitive unit is zero only in the absence of an externally applied stress (Figure 2.6).

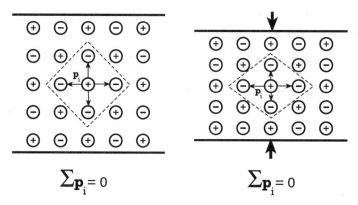

Figure 2.5 Illustration of the vanishing dipole in a hypothetical two-dimensional lattice. A crystal possessing a center of symmetry cannot be piezoelectric because the dipoles, p_i, within the primitive unit always cancel each other out. Hence, there is no net polarization within the crystal. An externally applied stress does not alter the center of symmetry. Adapted from Middelhoek and Audet [12].

The Sandbox: Materials for MEMS

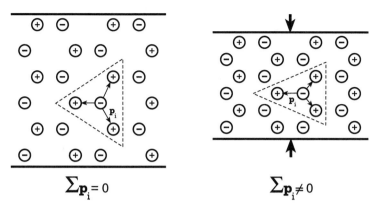

Figure 2.6 Illustration of the piezoelectric effect in a two-dimensional crystal. The net electric dipole within the primitive unit of an ionic crystal lacking a center of symmetry does not vanish when external stress is applied. This is the physical origin of piezoelectricity. Adapted from Middelhoek and Audet [12].

Straining the crystal, however, shifts the relative positions of the positive and negative charges giving rise to an electric dipole within the primitive unit and a net polarization across the crystal. Conversely, the internal electric dipoles realign themselves in response to an externally applied electric field causing the atoms to displace, and resulting in a measurable crystal deformation. When the temperature exceeds a critical value called the "Curie temperature," the crystal becomes cubic and loses its piezoelectric characteristics.

The piezoelectric effect is described in terms of piezoelectric charge coefficients, d_{3n} (in units of C/N), that relate the static voltage to displacement or applied force (Table 2.5). If a voltage, V_a, is applied across the thickness of a piezoelectric crystal (Figure 2.7), the displacements ΔL, ΔW, and Δt along the length, width, and thickness directions, respectively, are given by:

$$\Delta L = d_{31} \cdot V_a \cdot L/t \qquad \Delta W = d_{31} \cdot V_a \cdot W/t \qquad \Delta t = d_{33} \cdot V_a$$

where L and W are the length and width of the plate, respectively, and t is the thickness or separation between the electrodes. Conversely, if a force, F, is applied along any of the length, width, or thickness directions, a measured voltage, V_m, across the electrodes (in the thickness direction) is given in each of the three cases, respectively, by:

Table 2.5
Piezoelectric Coefficients and Other Relevant Properties
for a Selected List of Piezoelectric Materials

Material	Piezoelectric Constant (d_{3n}) (10^{-12} C/N)	Relative Permittivity (ε)	Density (g/cm^3)	Young's Modulus (GPa)	Acoustic Impedance (10^6 kg/m$^2 \cdot$ s)
Quartz	$d_{33} = 2.31$	4.5	2.65	107	15
Polyvinyledene-fluoride (PVDF)	$d_{31} = 23$ $d_{33} = -33$	12	1.78	3	2.7
LiNbO$_3$	$D_{31} = -4$, $d_{33} = 23$	28	4.6	245	34
BaTiO3	$d_{31} = 78$, $d_{33} = 190$	1,700	5.7		30
PZT	$D_{31} = -171$, $d_{33} = 370$	1,700	7.7	53	30
ZnO	$d_{31} = 5.2$, $d_{33} = 246$	1,400	5.7	123	33

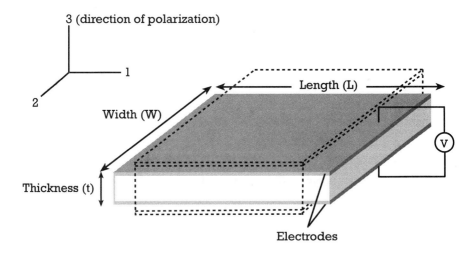

Figure 2.7 An illustration of the piezoelectric effect on a crystalline plate. An applied voltage across the electrodes results in dimensional changes in all three axes. Conversely, an applied force in any of three directions gives rise to a measurable voltage across the electrodes.

The Sandbox: Materials for MEMS

$$V_m = d_{31} \cdot F/(\varepsilon \cdot L) \qquad V_m = d_{31} \cdot F/(\varepsilon \cdot W) \qquad V_m = d_{33} \cdot F \cdot t/(\varepsilon \cdot L \cdot W)$$

where ε is the dielectric permittivity of the material. The reversibility between strain and voltage makes piezoelectric materials ideal for both sensing and actuation. Further detailed reading on piezoelectricity may be found in Cady [14] and Zelenka [15].

Quartz is a widely used stand-alone piezoelectric material, but there are no available methods to deposit crystalline quartz as a thin film over silicon substrates. Piezoelectric ceramics are also common. Lithium niobate ($LiNbO_3$) and barium titanate ($BaTiO_3$) are two well-known examples, but they are also difficult to deposit as thin films. Piezoelectric materials that can be deposited as thin film with relative ease are lead zirconate titanate (PZT)—a ceramic based on solid solutions of lead zirconate ($PbZrO_3$) and lead titanate ($PbTiO_3$)—zinc oxide (ZnO), and polyvinylidenefluoride (PVDF). Zinc oxide is typically sputtered and PZT can be either sputtered or deposited in a sol-gel process. The next chapter will describe the deposition processes in more detail. PVDF is a polymer that can be spun on, but it must be polarized by the application of a large electrical field across it, (poling), in order to exhibit a piezoelectrical behavior.

Thermoelectricity

Interactions between electricity and temperature are common and were the subject of extensive studies in the 19th century, though the underlying theory was not put in place until early in the 20th century by Boltzmann. In the absence of a magnetic field, there are three distinct thermoelectric effects: The Seebeck, Peltier, and Thomson effects [16]. The Seebeck effect is the most frequently used, for example, in thermocouples for the measurement of temperature differences. The Peltier effect is used to make thermoelectric coolers and refrigerators. The Thomson effect is less known and uncommon in daily applications. In the Peltier effect, a current flow across a junction of two dissimilar materials causes a heat flux, thus cooling one side and heating the other. Mobile wet bars with Peltier refrigerators were touted in the 1950s as the newest innovation in home appliances, but their economic viability was quickly jeopardized by the poor energy conversion efficiency. Today, Peltier devices are made of *n*-type and *p*-type bismuth telluride elements, and are used to cool high-performance microprocessors, laser diodes, and infrared sensors. Peltier devices have proven difficult to implement as micromachined thin film structures.

In the Seebeck effect, named for the scientist who made the discovery in 1822, a temperature gradient across an element gives rise to a measurable electric field that tends to oppose the charge flow (or electric current) resulting from the temperature imbalance. The measured voltage is, to first order, proportional to the temperature difference, with the proportionality constant known as the Seebeck coefficient. While, in theory, a single material is sufficient to measure temperature, in practice, thermocouples employ a junction of two dissimilar materials. The measurable voltage at the leads, ΔV, is the sum of voltages across both legs of the thermocouple (Figure 2.8). Therefore,

$$\Delta V = \alpha_1 \cdot (T_{cold} - T_{hot}) + \alpha_2 \cdot (T_{hot} - T_{cold}) = (\alpha_2 - \alpha_1) \cdot (T_{hot} - T_{cold})$$

where α_1 and α_2 are the Seebeck coefficients of materials 1 and 2, and T_{hot} and T_{cold} are the temperatures of the hot and cold sides of the thermocouple, respectively. Alternately, one may use this effect to generate electrical power by maintaining a temperature difference across a junction (Table 2.6).

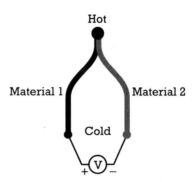

Figure 2.8 The basic structure of a thermocouple using the Seebeck effect. The measured voltage is proportional to the difference in temperature. Thermocouples can be readily implemented on silicon substrates using combinations of thin metal films or polysilicon.

The Sandbox: Materials for MEMS

Table 2.6
The Seebeck Coefficients Relative to Platinum for Selected Metals and for n- and p-Type Polysilicon.*

	μV/K		μV/K
Bi	−73.4	Ag	7.4
Ni	−14.8	Cu	7.6
Pa	−5.7	Zn	7.6
Pt	0	Au	7.8
Ta	3.3	W	11.2
Al	4.2	Mo	14.5
Sn	4.2	n-poly (30 Ω/□)	−100
Mg	4.4	n-poly (2600 Ω/□)	−450
Ir	6.5	p-poly (400 Ω/□)	270

* The sheet resistance is given for the 0.38-μm-thick polysilicon films. Polysilicon is an attractive material for the fabrication of thermocouples and thermopiles because of its large Seebeck coefficient.

Summary

The choice of substrate materials for MEMS is very broad, but crystalline silicon is by far the most common. Complementing silicon are a host of materials that can be deposited as thin films. These include polysilicon, amorphous silicon, silicon oxides and nitrides, glass, and organic polymers, as well as a host of metals. Crystallographic planes play an important role in the design and fabrication of silicon-based MEMS, and also affect some material properties of silicon. Three physical effects commonly used in the operation of micromachined sensors and actuators were introduced: Piezoresistivity, piezoelectricity, and thermoelectricity.

References

[1] Ashcroft, N. W. and N. D. Mermin, *Solid State Physics*, Philadelphia, PA: Saunders College, 1976, pp. 91–93.

[2] Kamins, T., *Polycrystalline Silicon for Integrated Circuits*, Boston, MA: Kluwer Academic Publishers, 1988.

[3] Bustillo, J. M., R. T. Howe, and R. S. Muller, "Surface Micromachining for Microelectromechanical Systems," in *Integrated Sensors, Microactuators, &*

Microsystems (MEMS), K. D. Wise (ed.), Proceedings of the IEEE, Vol. 86, No. 8, Aug. 1998, pp. 1559–1561.

[4] *Properties of Silicon*, EMIS Datareviews Series, No. 4, Inspec, New York, NY: IEE, 1988.

[5] Lorenz, H., M. Despont, N. Fahrni, N. LaBianca, P. Renaud, and P. Vettiger, "SU-8: A Low-Cost Negative Resist for MEMS," *Journal of Micromechanics and Microengineering*, Vol. 7, No. 3, Sept. 1997, pp. 121–124.

[6] Gutierrez Monreal, J., and C. M. Mari, "The Use of Polymer Materials as Sensitive Elements in Physical and Chemical Sensors," *Sensors and Actuators*, Vol. 12, 1987, pp. 129–144.

[7] Mehregany, M., C. A. Zorman, N. Rajan, and C. H. Wu, "Silicon Carbide MEMS for Harsh Environments," in *Integrated Sensors, Microactuators, & Microsystems (MEMS)*, pp. 1594–1610, K. D. Wise (ed.), Proceedings of the IEEE, Vol. 86, No. 8, Aug. 1998.

[8] Rogers, C., "Intelligent Materials," *Scientific American*, Vol. 273, No. 3, Sept. 1995, pp. 154–157.

[9] Smith, C. S., "Piezoresistive Effect in Germanium and Silicon," *Physics Review*, Vol. 94, 1954, pp. 42–49.

[10] Gieles, C. M., "Subminiature Silicon Pressure Sensor Transducer," *Digest IEEE International Solid-State Circuits Conference*, Philadelphia, PA, Feb. 19–21, 1969, pp. 108–109.

[11] Kanda, Y., "A Graphical Representation of the Piezoresistive Coefficients in Silicon," *IEEE Transactions on Electron Devices*, Vol. ED-29, No. 1, 1982, pp. 64–70.

[12] Middelhoek, S., and S. A. Audet, *Silicon Sensors*, San Diego, CA: Academic Press, 1989.

[13] Curie, P., and J. Curie, "Development by Pressure of Polar Electricity in Hemihedral Crystals with Inclined Faces," *Bull. Soc. Min. de France*, Vol. 3, 1880, p. 90.

[14] Cady, W. G., *Piezoelectricity*, New York, NY: Dover, 1964.

[15] Zelenka, J., *Piezoelectric Resonators and Their Applications*, Amsterdam, The Netherlands: Elsevier, 1986.

[16] MacDonald, D. K. C., *Thermoelectricity: An Introduction to the Principles*, New York, NY: Wiley, 1962.

Selected bibliography

Electrical Resistivity Handbook, G. T. Dyos and T. Farrell (eds.), London, England: Peter Pereginus, 1992.

Kittel, C., *Introduction to Solid State Physics*, 6th edition, New York, NY: Wiley, 1986.

Properties of Silicon, EMIS Datareviews Series, No. 4, Inspec, New York, NY: IEE, 1988.

Semiconductor Sensors, S. M. Sze (ed.), New York, NY: Wiley, 1994.

Sze, S. M., *Physics of Semiconductor Devices*, 2nd edition, New York, NY: Wiley, 1981.

CHAPTER 3

Contents

Basic process tools

Advanced process tools

Combining the tools—examples of commercial processes

Summary

The Toolbox: Processes for Micromachining

You will have to brace yourselves for this—not because it is difficult to understand, but because it is absolutely ridiculous: All we do is draw arrows on a piece of paper—that's all!

Richard Feynman, explaining the Theory of Quantum Electrodynamics. From the Alix G. Mautner Memorial Lectures, UCLA, 1983.

This chapter presents methods used in the fabrication of MEMS. Many are largely borrowed from the integrated circuit industry, in addition to a few others developed specifically for silicon micromachining. There is no doubt that the use of process equipment and the corresponding vast portfolio of fabrication processes developed initially for the semiconductor industry has given the burgeoning MEMS industry the impetus it needs to overcome the massive infrastructure requirements. For example, lithographic tools used in micromachining are often from previous generations of equipment designed for the fabrication of electronic integrated circuits. The equipment's performance is sufficient to meet the requirements of micromachining, but its price is substantially discounted. A few specialized processes such as anisotropic chemical wet etching, wafer bonding, deep reactive ion etching or sacrificial etching emerged over the years within the MEMS community, and remained limited to micromachining in their application.

From a simplistic perspective, micromachining bears a similarity to conventional machining in the sense that the objective is to precisely define arbitrary features in a block of material. However, there are distinct differences. Micromachining is a parallel (batch) process in which hundreds or possibly thousands of identical elements are fabricated simultaneously on the same wafer. Moreover, the minimum feature dimension is on the order of one micrometer, about a factor of 25 times smaller than what can be achieved using conventional machining.

Fundamentally, silicon micromachining combines adding layers of material over a silicon wafer with etching (in the sense of selectively removing material) precise patterns in these layers or the underlying substrate. The implementation is based on a broad portfolio of fabrication processes including material deposition, patterning, and etching techniques. Lithography plays a significant role in the delineation of accurate and precise patterns. These are the tools of MEMS (Figure 3.1).

We divide the toolbox into two major categories, basic and advanced. The basic process tools are well-established methods and are usually available at major foundry facilities; the advanced process tools are unique in their nature, and are normally limited to a few specialized facilities. For example, very few sites offer LIGA[1], a micromachining process using electroplating and molding.

Basic process tools

Epitaxy, sputtering, evaporation, chemical vapor deposition, and spin-on methods are common techniques used to deposit uniform layers of silicon, metals, insulators, or polymers. Lithography is a photographic process for printing images onto a layer of photosensitive polymer (photoresist) that is subsequently used as a protective mask against etching.

---

1. LIGA is a German acronym for "Lithographie, Galvanoformung und Abformung" meaning lithography, electroplating, and molding.

The Toolbox: Processes for Micromachining

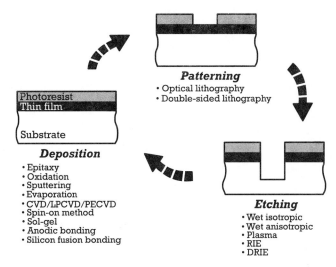

Figure 3.1 Illustration of the basic process flow in micromachining: layers are deposited; photoresist is lithographically patterned, and then used as a mask to etch the underlying materials. The process repeats until completion of the microstructure.

Wet and dry etching, including deep reactive ion etching, form the essential process base to selectively remove material. The following sections describe the fundamentals of each of the basic process tools.

Epitaxy

Epitaxy is a common deposition method to grow a crystalline silicon layer over a silicon wafer, but with a differing dopant type and concentration. The epitaxial layer is typically 1- to 20-μm thick. It exhibits the same crystal orientation as the underlying crystalline substrate, except when grown over an amorphous material, for example a layer of silicon dioxide, it is polycrystalline. Epitaxy is a widely used step in the fabrication of CMOS circuits, and has proven efficient in forming wafer-scale p-n junctions for controlled electrochemical etching (described later).

The growth occurs in a vapor-phase chemical deposition reactor from the dissociation at high temperature (> 800° C) of a silicon source gas. Common silicon sources are silane (SiH_4), silicon dichlorosilane (SiH_2Cl_2), or silicon tetrachloride ($SiCl_4$). Nominal growth rates vary between 0.2 and 1.5 μm/min depending on the source gas and the growth temperature. Impurity dopants are simultaneously incorporated during growth by the dissociation of a dopant source gas in the same reactor. Arsine (AsH_3) and phosphine (PH_3), two extremely toxic gases, are used for arsenic and phosphorous (n-type) doping, respectively; diborane (B_2H_6) is used for boron (p-type) doping.

Epitaxy can be used to grow crystal silicon on other types of crystalline substrates such as sapphire (Al_2O_3). The process is called heteroepitaxy to indicate the difference in materials. Silicon-on-sapphire (SOS) wafers are available from a number of vendors, and are effective in applications where an insulating or a transparent substrate is required. The lattice mismatch between the sapphire and silicon crystals limits the thickness of the silicon to about one-micrometer. Thicker silicon films suffer from high defect densities and degraded electronic performance.

Oxidation

High-quality silicon dioxide is obtained by oxidizing silicon in either dry oxygen or in steam at elevated temperatures (850–1150° C). Oxidation mechanisms have been extensively studied and are well understood. Charts showing final oxide thickness as a function of temperature, oxidizing environment, and time are widely available [1].

Thermal oxidation of silicon generates compressive stress in the silicon dioxide film. There are two reasons for the stress: Silicon dioxide molecules take more volume than silicon atoms, and there is a mismatch between the coefficients of thermal expansion of silicon and silicon dioxide. The compressive stress depends on the total thickness of the silicon dioxide layer, and can reach hundreds of MPa. As a result, thermally grown oxide films thicker than one micrometer can cause bowing of the underlying substrate. Moreover, freestanding membranes and suspended cantilevers made of thermally grown silicon oxide tend to warp or curl.

Sputter deposition

In sputter deposition, a target object made of a material to be deposited is physically bombarded by a flux of inert ions (e.g., argon, helium) in a vacuum chamber. Material particles from the target are ejected and deposited on the wafer. There are three general classes of sputter tools differing by the ion excitation mechanism. In *DC glow discharge*, the inert ions are accelerated in a DC field between the target and the wafer. In *planar*

RF, the target and the wafer form two parallel plates with RF excitation applied to the target. Both DC and RF planar sputter methods work well for the deposition of insulating materials such as glass. In *planar* and *cylindrical magnetron* (or S-gun), an externally applied magnetic field increases the ion density near the target, thus raising the deposition rates. For certain materials such as aluminum, the deposition rate can be as high as 1 μm/min.

Sputtering is a favored method in the MEMS community for the deposition at low temperatures (< 150° C) of thin metal films such as aluminum, titanium, chromium, platinum, and palladium, as well as amorphous silicon and insulators including glass and piezoelectric ceramics (e.g., PZT, ZnO). The directional randomness of the sputtering process, provided that the target size is larger than the wafer, results in good "step coverage"—the uniformity of the thin film over a geometrical step—though some thinning occurs near corners.

The deposited film has a very fine granular structure and is frequently under stress [2]. The stress levels normally vary with the chamber pressure during deposition from compressive at low pressures (0.1–1 Pa) to tensile at high pressures (1–10 Pa). The transition between the compressive and tensile regimes is often sharp (over a few tenths of Pa) making the crossover, an ideal point for zero-stress deposition, difficult to control. Raising the substrate temperature typically results in a decrease in stress, especially for metals with low melting points such as aluminum.

Evaporation

Evaporation involves the local heating of a target material to a sufficiently high temperature in order to generate a vapor that condenses on a substrate. Nearly any material (e.g., Al, Si, Ti, Mo, glass, Al_2O_3 ... and so on), including many high melting point refractory metals (W, Au, Cr, Pd, Pt), can be evaporated provided it has a vapor pressure above the background pressure (0.1–1 Pa), and that the carrier in which the target is contained is itself not evaporated—the carrier is usually made of tungsten.

Target heating is accomplished either resistively by passing an electrical current through a filament made of the desired target material, or by scanning an electron beam over the target. In the latter case, electrons emitted from a hot filament are accelerated in a 10 kV-potential before striking and melting the target. Resistive evaporation is simple, but can result in spreading impurities or other contaminants present in the filament. The small size of the filament also limits the thickness of the deposited film. Electron beam evaporation, by contrast, can provide better quality films and higher deposition rates (50–500 nm/min), but the deposition system is more complex, requiring water cooling of the target, and shielding from x-rays generated when the energetic electrons strike the target. Furthermore, radiation that penetrates the surface of the silicon substrate during the deposition process can damage the crystal and degrade the characteristics of electronic circuits.

Evaporation is a directional deposition process whereby the vast majority of material particles are deposited at a specific angle to the substrate, resulting in poor step coverage and leaving corners and sidewalls exposed. This is generally an undesirable effect if thin film continuity is desired, for example, the metal is an electrical interconnect. Rotating the substrate during deposition reduces the effect. However, in some cases, this shadowing can be used deliberately to selectively deposit material on one side of a step or a trench but not the other.

Thin films deposited by evaporation exhibit high tensile stresses, increasing with higher material melting point. Evaporated tungsten or nickel films, for example, can have stress in excess of 500 MPa, sufficient to cause curling or even peeling. Similar to sputtering, raising the deposition temperature of the substrate tends to reduce the stress in the thin film.

Chemical vapor deposition

Chemical vapor deposition (CVD) works on the principle of initiating a chemical reaction in a vacuum chamber, resulting in the deposition of a reacted species on a heated substrate. In contrast to sputtering, CVD is a high-temperature process with typical deposition temperatures above 300° C. The field of CVD has grown substantially, driven by the demand within the semiconductor industry for high-quality thin dielectric and metal films for multilayer electrical interconnects. Common thin films deposited by CVD include polysilicon, silicon oxides and nitrides, tungsten, titanium, and tantalum as well as their nitrides, and most recently, copper and low permittivity dielectric insulators ($\varepsilon_r < 3$). The latter two are becoming workhorse materials for very high-speed electrical interconnects in integrated circuits. The deposition of polysilicon, and silicon oxides and nitrides is routine within the MEMS industry.

Chemical vapor deposition processes are categorized as *atmospheric pressure* (referred to as CVD), or *low pressure* (LPCVD), or *plasma[2]-enhanced* (PECVD), which also encompasses high-density plasma (HDP-CVD). CVD and LPCVD methods operate at rather elevated temperatures (500 to 800° C). In PECVD and HDP-CVD, the substrate temperature is typically near 300° C, though the plasma deposition of silicon nitrides at room temperature is feasible. The effect of deposition parameters on the characteristics of the thin film is significant, especially for silicon oxides and nitrides. Substrate temperature, gas flows, presence of dopants, and pressure are important process variables for CVD and LPCVD. Power and plasma excitation RF-frequency are also important for PECVD.

Deposition of polysilicon

Chemical vapor deposition processes allow the deposition of polysilicon as a thin film on a silicon substrate. The film thickness can range between a few tens of nanometers to several micrometers. Structures with multiple layers of polysilicon deposited one at a time are feasible. The ease of depositing polysilicon, a material sharing many of the properties of bulk silicon, makes it an extremely attractive material in surface micromachining.

Polysilicon is deposited by the pyrolysis of silane (SiH_4) to silicon and hydrogen in a LPCVD reactor. Deposition from silane in a low temperature PECVD reactor is also possible, but results in amorphous silicon. The deposition temperature in LPCVD, typically between 550 and 700° C, affects the granular structure of the film. Below 600° C the thin film is completely amorphous; above 630° C it exhibits a crystalline grain structure. While the polycrystalline film contains grains of all orientations, the preferred orientation is {110}, changing to {100} when the deposition temperature exceeds 650° C. The deposition rate varies from approximately 10 nm/min at 630° C up to 70 nm/min at 700° C. Partial pressure and flow rate of the silane gas also affect the deposition rate.

---

2. Energetic electrons excited in a high-frequency electromagnetic field collide with gas molecules to form ions and reactive neutral species. The mixture of electrons, ions, and neutrals is called a plasma, and constitutes a phase of matter distinct from solids, liquids, or gases. Plasma-phase operation increases the density of ions and neutral species that can participate in a chemical reaction, whether it is deposition or etching, and thus can accelerate the reaction rate.

Generally speaking, CVD polysilicon films conform well to the underlying topography on the wafer, and show good step coverage. In deep trenches with aspect ratios (ratio of depth to width) in excess of 10, some thinning of the film occurs on the sidewalls, but that has not limited the use of polysilicon to fill trenches as deep as 100 μm.

Polysilicon can be doped during deposition—known as *in situ* doping—by introducing dopant source gases, in particular arsine (AsH_3) or phosphine (PH_3) for *n*-type doping, and diborane (B_2H_6) for *p*-type doping. Arsine and phosphine decrease the deposition rate, whereas diborane increases it. The dopant concentration in *in situ*-doped films is normally very high ($\sim 10^{20}$ cm^{-3}), but the film resistivity remains in the range of 1 to 10 m$\Omega \cdot$cm because of the low mobility of electrons or holes.

Intrinsic stresses in as-deposited doped polysilicon films are large (> 500 MPa), and can result in the warping or curling of released micromechanical structures made of such layers. Films deposited below 600° C exhibit tensile stresses, whereas compressive stresses are observed at higher temperatures. Annealing at 900° C or above causes stress relaxation through structural changes in grain boundaries, and a reduction in stress to levels (< 50 MPa) generally deemed acceptable for micromachined structures.

Deposition of silicon dioxide

Silicon oxide is deposited below 500° C by reacting silane and oxygen in a CVD, LPCVD, or PECVD reactor. The optional addition of phosphine or diborane dopes the silicon oxide with phosphorus or boron, respectively. Films doped with phosphorus are often referred to as phosphosilicate glass (PSG); those doped with phosphorus and boron are known as borophosphosilicate glass (BPSG), or simply low-temperature oxide (LTO). At temperatures near 1000° C, both PSG and BPSG soften and flow to conform to the underlying surface topography, and to improve step coverage. LTO films make good passivation coatings over aluminum, but the deposition temperature must remain below ~350° C to prevent degradation of the metal.

Silicon dioxide can also be deposited at temperatures between 650° C and 750° C in a LPCVD reactor by the pyrolysis of tetraethoxysilane [$Si(OC_2H_5)_4$], also known as TEOS. Silicon dioxide layers deposited from a TEOS source exhibit excellent uniformity and step coverage, but the high temperature process precludes their use over aluminum.

A third but less common method to deposit silicon dioxide involves reacting dichlorosilane ($SiCl_2H_2$) with nitrous oxide (N_2O) in a LPCVD reactor at temperatures near 900° C. Film properties and uniformity are excellent, but its use is limited to depositing insulating layers over polysilicon. Oxide doping is very difficult because of the high deposition temperature.

As is the case for the CVD of polysilicon, deposition rates for silicon dioxide increase with temperature. A typical LTO deposition rate at atmospheric pressure is 150 nm/min at 450° C; the deposition rates using TEOS vary between 5 nm/min at 650° C up to 50 nm/min at 750° C.

Deposited silicon dioxide films are amorphous with a structure similar to fused silica. Heat treatment at elevated temperatures (600–1000° C) results in an increase in density accompanied by a reduction in film thickness, but no change in the amorphous structure. This process is called densification.

Silicon dioxide deposited using CVD methods is very useful as a dielectric insulator between layers of metal, or as a sacrificial layer (etched using hydrofluoric acid) in surface micromachining. However, its electric properties are inferior to those of thermally grown silicon dioxide. For example, the dielectric strength of CVD silicon oxides can be half that of thermally grown silicon dioxide. It is no coincidence that gate insulators for CMOS transistors are made of the latter type. In general, CVD silicon oxides are under compressive stress (100–300 MPa). The stress cannot be controlled except when PECVD is used.

Deposition of silicon nitrides

Silicon nitride is common in the semiconductor industry for the passivation of electronic devices because it forms an excellent protective barrier against the diffusion of water and sodium ions. In micromachining, silicon nitride films are effective as masks for the selective etching of silicon in alkaline solutions such as potassium hydroxide.

Stoichiometric silicon nitride (Si_3N_4) is deposited at atmospheric pressure by reacting silane (SiH_4) and ammonia (NH_3), or at low pressure by reacting dichlorosilane ($SiCl_2H_2$) and ammonia. The deposition temperature for either method is between 700° and 900° C. Both reactions generate hydrogen as a byproduct, which is incorporated into the deposited film. CVD and LPCVD silicon nitride films generally exhibit large tensile stresses approaching 1000 MPa. However, if the silicon nitride is silicon-rich, i.e., there is an excess of silicon in the film, then the stress is below 100 MPa, a level considered acceptable for most micromachining applications.

For deposition below 400° C, nonstoichiometric silicon nitride (Si_xN_y) is obtained by reacting silane with ammonia in a plasma-enhanced deposition (PECVD) chamber. Hydrogen is also a byproduct of this reaction, and is incorporated in elevated concentrations (20–25%) in the film. The refractive index is an indirect measure of impurity content and overall quality of the silicon nitride film. It ranges between 1.8 and 2.5 for PECVD films—the refractive index for stoichiometric LPCVD silicon nitride is 2.01. A high value in the range is indicative of excess silicon, and a low value generally represents an excess of oxygen.

One of the key advantages of PECVD nitride is the ability to control stress during deposition. Silicon nitride deposited at a plasma excitation frequency of 13.56 MHz exhibits tensile stress of about 400 MPa, whereas a film deposited at a frequency of 50 kHz has a compressive stress of 200 MPa. By alternating frequencies during deposition, one may obtain nearly stress-free films.

Spin-on methods

Spin-on is a simple process to put down layers of dielectric insulators and organic materials. Unlike the methods described earlier, the equipment is simple requiring a variable speed-spinning table with appropriate safety screens. A nozzle dispenses the material as a liquid solution in the center of the wafer. Spinning the substrate at high speeds (500 to 5000 rpm) rapidly spreads the material in a uniform manner.

Photoresist and polyimides are common organic materials that can be spun on a wafer with thicknesses typically between 0.5 and 20 μm, though some special purpose resists, such as the epoxy-based SU-8, can reach a thickness of 100 μm. The organic polymer is normally in suspension in a solvent solution. Subsequent baking or exposure to ultraviolet radiation causes the solvent to evaporate, and cures the film.

Thick (5–100 μm) spin-on glass (SOG) has the ability to uniformly coat surfaces and smooth out underlying topographical variations, effectively "planarizing" surface features. Thin (0.1–0.5 μm) SOG was heavily investigated in the integrated circuit industry as an interlayer dielectric between metals for high-speed electrical interconnects; however, its electrical properties are considered poor compared with native or CVD silicon oxides. Spin-on glass is commercially available in different forms of polymers, commonly siloxane- or silicate-based. The latter type allows water absorption into the film, resulting in a higher relative dielectric constant and a tendency to crack. After deposition, the layer is typically densified at a temperature between 300° and 500° C. Measured film stress is approximately 200 MPa in tension, but decreases substantially with increasing anneal temperatures.

Lithography

Lithography involves three sequential steps:

- Application of photoresist (or resist) which is a photosensitive emulsion layer;
- Optical exposure to print an image of the mask onto the resist;
- Immersion in an aqueous developer solution to dissolve the exposed resist and render visible the latent image.

The mask itself consists of a patterned opaque chromium layer on a transparent glass substrate. The pattern layout is generated using a computer-aided design (CAD) tool, and transferred into the thin chromium layer at a specialized mask-making facility. A complete microfabrication process frequently involves several lithographic operations.

Positive photoresist is an organic resin material containing a "sensitizer." It is spin-coated on the wafer with a typical thickness between 0.5 μm and 10 μm. As mentioned earlier, special types of resists can be spun to thicknesses of up to 100 μm, but the large thickness poses significant challenges to exposing and defining features below 25 μm in size. The sensitizer prevents the dissolution of unexposed resist during immersion in the developer solution. Exposure to light in the 200 to 450 nm range (ultraviolet to blue) breaks down the sensitizer, causing exposed regions to immediately dissolve in developer solution. The exact opposite process happens in *negative* resists—exposed areas remain and unexposed areas dissolve in the developer.

Optical exposure can be accomplished in one of three different modes: contact, proximity, or projection. In contact lithography, the mask touches the wafer. This normally shortens the life of the mask, and leaves undesired residue on the wafer and the mask. In proximity mode, the mask is brought to within 25–50 μm of the resist surface. In contrast, projection lithography projects an image of the mask onto the wafer through complex optics (Figure 3.2).

Resolution, defined as the minimum feature the optical system can resolve, is seldom a limitation for micromachining applications. For proximity systems it is limited by Fresnel diffraction to a minimum of about 5 μm, and in contact systems it is approximately 1 to 2 μm. For projection systems it is given by $0.5 \times \lambda/NA$, where λ is the wavelength (~ 400 nm)

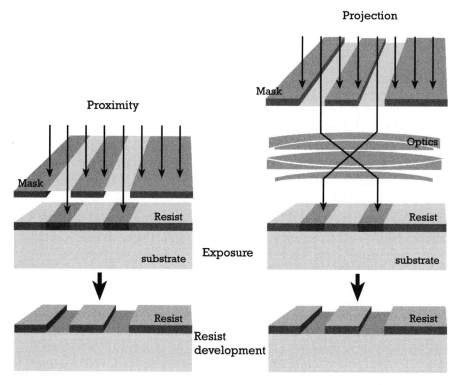

Figure 3.2 An illustration of proximity and projection lithography. In proximity mode, the mask is within 25–50 μm of the resist. Fresnel diffraction limits the resolution and minimum feature size to ~5 μm. In projection mode, complex optics image the mask onto the resist. The resolution is routinely better than one micrometer. Subsequent development delineates the features in the resist.

The Toolbox: Processes for Micromachining

and *NA* is the numerical aperture of the optics. Resolution in projection lithography is routinely better than one micrometer. Depth of focus, however, is a more severe constraint on lithography, especially in light of the need to expose thick resist, or accommodate geometrical height variations across the wafer. Depth of focus for contact and proximity systems is very poor, also limited by Fresnel diffraction. In projection systems, the image plane can be moved by adjusting the focus settings, but once it is fixed, the depth of focus about that plane is limited to $\pm 0.5 \times \lambda/NA^2$. In nearly all cases, depth of focus is at most a few microns.

Projection lithography is clearly a superior approach, but an optical projection system can cost significantly more than a proximity or contact system. Long-term cost of ownership plays a critical role in the decision to acquire a particular lithographic tool.

While resolution of most lithographic systems is not a limitation, lithography for MEMS can be challenging, depending on the nature of the application. Exposure of thick resist, topographical height variations, front-to-backside pattern alignment, and large fields of view are examples.

Thick resist

Patterned thick resist is normally used as a protective masking layer for the etching of deep structures, but in some instances, it is a template for the electroplating of metal microstructures. Coating substrates with thick resist is achieved either by multiple spin-coating applications (up to a total of 10 μm), or by spinning special viscous resist solutions at slower speeds (up to 100 μm). Maintaining thickness control and uniformity across the wafer becomes difficult with increasing resist thickness.

Exposing resist thicker than 5 μm often degrades the minimum resolvable feature size due to the limited depth of focus of the exposure tool—different planes within the resist will be imaged differently. The net result is a sloping of the resist profile in the exposed region. As a general guideline, the maximum aspect ratio (ratio of resist thickness to minimum feature dimension) is approximately three—in other words, the minimum achievable feature size is larger than one third of the resist thickness. This limitation may be overcome using special exposure methods, but their value in a manufacturing environment remains questionable.

Topographical height variations

Changes in topography on the surface of the wafer, such as deep cavities and trenches, are very common in MEMS, and pose challenges to both resist spinning and imaging. For cavities deeper than 10 μm, achieving uniformity becomes a tedious task because of thinning of the resist at convex corners and accumulation inside the cavity (Figure 3.3). Though resist spraying can be used to deposit rather thick (> 5 μm) layers, generally, process engineers frown upon the task of having to coat wafers with significant variations in height.

Exposing a pattern on a surface with height variations in excess of 10 μm is also a difficult task because of the limited depth of focus. Contact and proximity tools are unsuitable for this task, unless a significant loss of resolution is tolerable. But under certain circumstances where the number of height levels is limited (say, less than three), one may use a projection lithography tool to perform an exposure with a corresponding focus adjustment at each of these height levels. Naturally, this is costly because the number of masks and exposures increases linearly with the number of height levels.

Double-sided lithography

Often, lithographic patterns on both sides of a wafer need to be aligned with respect to each other with a high degree of accuracy. For example, the fabrication of a pressure sensor entails forming on the front side of the wafer piezoresistive sense elements that are aligned to the edges of a cavity on the back side of the wafer. Relative misalignments greater than 5 μm alter the sensitivity of the piezoresistive bridge to pressure, and create undesirable second-order effects. Different methods of front-to-

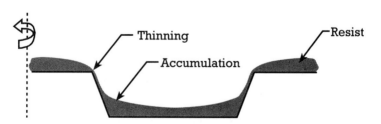

Figure 3.3 Undesirable effects of spin coating resist on a surface with severe topographical height variations. The resist is thin on corners and accumulates in the cavity.

backside alignment, also known as double-sided alignment, have been incorporated in commercially available tools. Wafers polished on both sides should be used to minimize light scattering.

Two companies, Karl Süss GmbH, of Munich, Germany, and Electronic Visions Company of Schärding, Austria, provide equipment capable of double-sided alignment and exposure. The operation of the Karl Süss MA-150 production-mode system uses a patented scheme to align crosshair marks on the mask to crosshair marks on the backside of the wafer. First, the alignment marks on the mechanically clamped mask are viewed by a set of dual objectives, and an image is electronically stored. The wafer is then loaded with the backside alignment marks facing the microscope objectives, and positioned so that these marks are aligned to the electronically stored image. After alignment, exposure of the mask onto the front side of the wafer is completed in proximity or contact mode. A typical registration error (or misalignment) is less than 2 μm (Figure 3.4).

Large field of view

The field of view is the extent of the area that is exposed at any one time on the wafer. In proximity and contact lithography, it covers the entire wafer. In projection systems, the field of view is often less than 2×2 cm^2. The entire wafer is exposed by stepping the small field of view across in a two-dimensional array, hence the "stepper" appellation. In some applications, the device structure may span dimensions exceeding the field of view. A remedy to this is called "field stitching" in which two or more different fields are exposed sequentially in juxtaposition.

Etching

Etch processes for MEMS fabrication deviate from traditional etch processes for the integrated circuit industry. While it has many of its underpinnings in science, etching for micromachining remains to a large extent an art. The objective is to selectively remove material using imaged photoresist as a masking template. The pattern can be etched directly into the silicon substrate, or into a thin film which in turn can be used as a mask for subsequent etches. For a successful etch, there must be sufficient selectivity between the masking material and the material being etched.

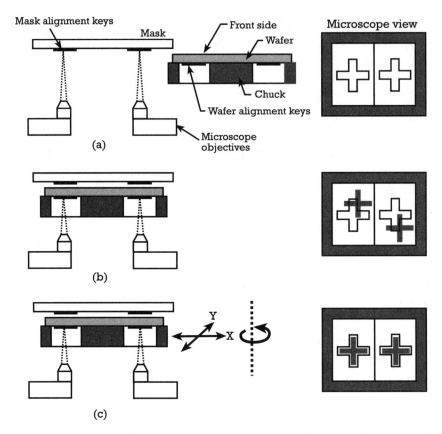

Figure 3.4 Double-sided alignment scheme for the Karl Süss MA-150 production mode system: (a) The image of mask alignment marks is electronically stored; (b) The alignment marks on the backside of the wafer are brought in focus; (c) The position of the wafer is adjusted by translation and rotation to align the marks to the stored image. The right-hand-side illustrates the view on the computer screen as the targets are brought into alignment. Adapted from product technical sheet (Karl Süss GmbH, Munich, Germany).

Etching thin films is relatively easier than etching bulk silicon. Table 3.1 provides a list of wet and dry (plasma phase) etchants commonly used for thin metal films and dielectric insulators.

Etching of silicon lies at the core of what is often termed "bulk micromachining." No ideal silicon etch method exists, leaving process engineers with a number of techniques, each suitable for some

The Toolbox: Processes for Micromachining

Table 3.1
Wet and Dry Etchants of Thin Metal Films and Dielectric Insulators.
Adapted from Williams and Muller [3].

	Wet Etchants (aqueous solutions)	Etch Rate (nm/min)	Dry Etching Gases (plasma phase)	Etch Rate (nm/min)
Silicon dioxide	HF	20–2,000	$CHF_3 + O_2$	50–150
	$HF:NH_4F$ (buffered HF)	100–500	$CHF_3 + CF_4$	250–500
Silicon nitride	H_3PO_4	5	SF_6	150–250
			$CHF_3 + CF_4$	100–150
Aluminum	$H_3PO_4:HNO_3:CH_3COOH$	660	$Cl_2 + SiCl_4$	100–150
	HF	5	$CHCl_3 + BCl_3$	200–600
Gold	KI	40		
Titanium	$HF:H_2O_2$	880	SF_6	100–150
Tungsten	H_2O_2	20–100	SF_6	100–150
	$K_3Fe(CN)_6:KOH:KH_2PO_4$	34		
Chromium	$Ce(NH_4)(NO_3)_6:HCl_4$	2	Cl_2	5
	HCl			
Organic layers	$H_2SO_4:H_2O_2$	> 1,000	O_2	35–3,500
	CH_3COCH_3 (acetone)	> 4,000		

applications but not others. Distinctions are made on the basis of isotropy, etch medium, and selectivity of the etch to other materials.

Isotropic etchants etch uniformly in all directions, resulting in rounded cross-sectional features. In contrast, anisotropic etchants etch in one direction preferentially over others, resulting in trenches or cavities delineated by flat and well-defined surfaces; these need not be perpendicular to the surface of the wafer. The etch medium (wet vs. dry) plays a role in selecting a suitable method. Wet etchants in aqueous solution offer the advantage of low-cost batch fabrication—usually 20 to 25 wafers can be etched simultaneously—and can be either of the isotropic or anisotropic type. Dry etching involves the use of reactant gases in a low-pressure plasma. The equipment is specialized, and requires the plumbing of ultra-clean pipes to bring high-purity reactant gases into the vacuum chamber (Figure 3.5).

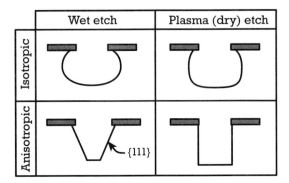

Figure 3.5 Schematic illustration of cross-sectional trench profiles resulting from four different types of etch methods.

Isotropic wet etching

The most common of isotropic wet etchants is "HNA," also known as "poly-etch" because of its use in the early days of the integrated circuit industry as an etchant for polysilicon. It is a mixture of hydrofluoric (HF), nitric (HNO_3), and acetic (CH_3COOH) acids (Table 3.2). Nitric acid oxidizes silicon which is removed by the hydrofluoric acid. The etch rate of silicon can vary between 1 and 5 μm/min, depending on the proportion of the acids in the mixture. Stirring can further increase the etch rate to 20 μm/min. Etch depth and surface uniformity are normally difficult to control.

Anisotropic wet etching

The list of anisotropic wet etchants includes the hydroxides of alkali metals (e.g., NaOH, KOH, CsOH ... etc.), simple and quaternary ammonium hydroxides, (e.g., NH_4OH, $(CH_3)_4NOH$), and ethylenediamine mixed with pyrochatechol in water (EDP) (Table 3.2).

Of the hydroxides of alkali metals, potassium hydroxide (KOH) is by far the most common. KOH etches {111} planes at a rate 100 times slower than it etches {100} planes [4]. This feature is routinely used to make V-shaped grooves and trenches that are precisely delineated by {111} crystallographic planes. The overall reaction consists of the oxidation of silicon followed by a reduction step:

$$Si + 2OH^- \rightarrow Si(OH)_2^{++} + 4e^- \text{ (oxidation)}$$

Table 3.2
Liquid and Gas Phase Etchants of Silicon. Adapted from Kovacs et al. [5].

	HF:HNO$_3$: CH$_3$COOH	KOH	Ethylene-diamine pyrochatechol (EDP)	(CH$_3$)$_4$NOH (TMAH)	SF$_6$	SF$_6$/C$_4$F$_8$ (DRIE)
Etch type	Wet	Wet	Wet	Wet	Plasma	Plasma
Typical formulation	250 ml HF, 500 ml HNO$_3$, 800 ml CH$_3$COOH	40 to 50 wt%	750 ml Ethylene-diamine, 120 g Pyrochatechol, 100 ml water	20 to 25 wt%		
Anisotropic	No	Yes	Yes	Yes	Varies	Yes
Temperature	25° C	70–90° C	115° C	90° C	0–100° C	20–80° C
Etch rate (μm/min)	1 to 20	0.5 to 2	0.75	0.5 to 1.5	0.1 to 0.5	1 to 3
{111}/{100} selectivity	None	100:1	35:1	50:1	None	None
Nitride etch (nm/min)	Low	<1	0.1	<0.1	200	200
SiO$_2$ etch (nm/min)	10–30	10	0.2	<0.1	10	10
p^{++} etch stop	No	Yes	Yes	Yes	No	No

$$\text{Si(OH)}_2{}^{++} + 4e^- + 4H_2O \rightarrow \text{Si(OH)}_6{}^- + 2H_2 \text{ (reduction)}$$

A charge transfer of four electrons occurs during the reaction.

There is little consensus on the origin of the selectivity to {111} crystallographic planes. Proposals made throughout the literature attribute the anisotropy to the lower bond density—and hence lower electron concentration—along {111} planes. Others believe that {111} planes oxidize quickly, and are protected during the etch with a thin layer of oxide.

KOH and alkaline etchants are also selective to heavily doped p-type (p^{++}) silicon, making common the use of p^{++} doping as an etch stop [4]. The etch rate of silicon in KOH solutions is approximately 0.5 to 2 μm/min depending on the temperature and the concentration of KOH, but it drops by a factor of 500 in p^{++} silicon with a dopant concentration above

1×10^{20} cm^{-3}. It is believed that the heavy p-type doping deprives the chemical reaction of electrons critical to the oxidation of silicon.

Silicon nitride is an excellent masking material against etching in KOH. Silicon dioxide etches at 10 nm/min and is used as a masking layer for very short etches. Photoresist is readily etched in alkaline solutions, and is not suitable for masking silicon etches.

Alkali hydroxides are extremely corrosive; aluminum bond pads inadvertently exposed to KOH are quickly damaged. It is important to note that CMOS fabrication facilities are very reluctant to use such etchants, or even to accept wafers previously exposed to alkali hydroxides for fear of contamination of potassium or sodium, two ions detrimental to the operation of CMOS electronic circuitry.

In the category of ammonium hydroxides, tetramethyl ammonium hydroxide (TMAH) ((CH_3)$_4$NOH)) exhibits similar properties to KOH [6]. It etches {111} crystallographic planes 30 to 50 slower than {100} planes. The etch rate drops by a factor of 40 in heavily p-doped silicon ($\sim 1 \times 10^{20}$ cm^{-3}). A disadvantage of TMAH is the occasional formation of undesirable pyramidal hillocks at the bottom of the etched cavity. Both silicon dioxide and silicon nitride remain virtually unetched in TMAH, and hence can be used as masking layers. It is advisable to remove native silicon dioxide in hydrofluoric acid prior to etching in TMAH because a layer just a few nanometers thick is sufficient to protect the silicon surface from etching. TMAH normally attacks aluminum, but a special formulation containing silicon powder dissolved in solution significantly reduces the etch rate of aluminum [7]. This property is useful for the etching of silicon after the complete fabrication of CMOS circuits, without resorting to the masking of the aluminum bond pads.

EDP (Ethylenediamine pyrochatechol) is another wet etchant with selectivity to {111} planes and to heavily p-doped silicon. It is extremely hazardous and its vapors are carcinogenic, necessitating the use of completely enclosed reflux condensers. Silicon oxides and nitrides make excellent masking materials for EDP etching. Many metals, including gold, chromium, copper, and tantalum are also not attacked in EDP. The formulation given in Table 3.2 etches aluminum at an approximate rate of 20 μm/hr.

Etching using anisotropic aqueous solutions results in three-dimensional faceted structures formed by intersecting {111} planes with other crystallographic planes. The design of the masking pattern demands a visualization in three dimensions of the etch procession. To that end, etch computer simulation software, such as the program ACES™ available from the University of Illinois at Urbana-Champaign, are useful design tools.

The easiest structures to visualize are V-shaped cavities etched in {100}-oriented wafers. The etch front begins at the opening in the mask and proceeds in the <100> direction, which is the vertical direction in {100}-oriented substrates, creating a cavity with a flat bottom and slanted sides. The sides are {111} planes making 54.74° with respect to the horizontal {100} planes. The etch ultimately self limits on four equivalent but intersecting {111} planes, forming an inverted pyramid or V-shaped trench. Of course, this occurs only if the wafer is thicker than the projected etch depth. Timed etching from one side of the wafer is frequently used to form cavities or thin membranes. Funnel and oblique-shaped ports are also possible in {100} wafers by etching aligned patterns from both sides of the wafer, and allowing the two etch fronts to coalesce (Figure 3.6).

The shape of an etched trench in {110} wafers is radically different. In silicon {110} wafers, four of the eight equivalent {111} planes are perpendicular to the {110} wafer surface. The remaining four {111} planes are slanted at 35.26° with respect to the surface. The four vertical {111} planes intersect to form a parallelogram with an inside angle of 70.5°. A groove etched in {110} wafers has the appearance of a complex polygon delineated by six {111} planes, four vertical and two slanted. Etching in {110} wafers is useful to form trenches with vertical sidewalls, albeit not orthogonal to each other [8] (Figure 3.7).

While concave corners bounded by {111} planes remain intact during the etch, convex corners are immediately attacked. This is because any slight erosion of the convex corner exposes planes other than {111} planes, thus accelerating the etch. Consequently, a convex corner in the mask layout will get undercut during the etch; in other words, the etch front will proceed underneath the masking layer. In some instances, such as when a square island is desired, this effect becomes detrimental, and is compensated for by clever layout schemes called "corner compensation" [9]. Often, however, the effect is intentionally used to form beams suspended over cavities (Figures 3.8 and 3.9).

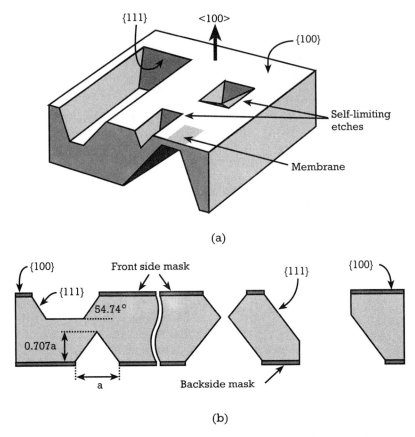

Figure 3.6 Illustration of the anisotropic etching of cavities in {100}-oriented silicon: (a) cavities, self-limiting pyramidal and V-shaped pits, and thin membranes; (b) etching from both sides of the wafer can yield a multitude of different shapes including funnel-shaped or oblique holes.

Electrochemical Etching

The relatively large etch rates of anisotropic wet etchants (> 0.5 μm/min) make it difficult to achieve uniform and controlled etch depths. Some applications, such as bulk micromachined pressure sensors, demand a thin (5 to 20 μm) silicon membrane with dimensional thickness control and uniformity of better than 0.2 μm, which is very difficult to achieve using timed etching. Instead, the thickness control is obtained by using a precisely grown epitaxial layer, and modulating the etch reaction with an externally applied electrical potential. This method is commonly referred to as "electrochemical etching" (ECE) [10,11]. An *n*-type epitaxial layer

The Toolbox: Processes for Micromachining 63

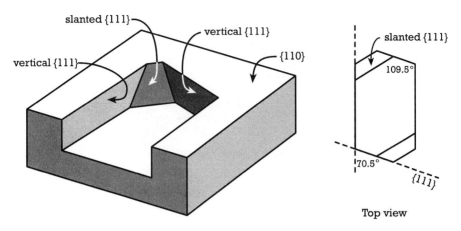

Figure 3.7 Illustration of the anisotropic etching in {110}-oriented silicon. Etched structures are delineated by four vertical {111} planes and two slanted {111} planes. The vertical {111} planes intersect at an angle of 70.5°.

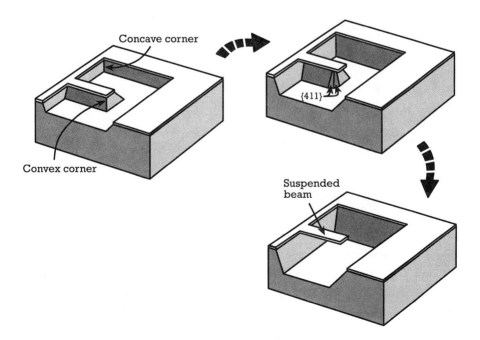

Figure 3.8 Illustration of the etching at convex corners and the formation of suspended beams.

Figure 3.9 Photograph of a thermally isolated RMS converter consisting of thermopiles on a silicon dioxide membrane. The anisotropic etch undercuts the silicon dioxide mask to form a suspended membrane. Courtesy of D. Jaeggi, Swiss Federal Institute of Technology, Zurich, Switzerland.

grown on a p-type wafer forms a p-n junction diode that allows electrical conduction only if the p-type side is at a voltage above the n-type; otherwise, no electrical current passes and the diode is said to be in "reverse bias." The applied potential is such that the p-n diode is in reverse bias, and the n-type epitaxial layer is above its passivation potential—the potential at which a thin passivating silicon dioxide layer forms—hence it is not etched. The p-type substrate is allowed to electrically float, and so it is etched. As soon as the p-type substrate is completely removed, the etch reaction comes to a halt at the junction, leaving a layer of n-type silicon with precise thickness (Figure 3.10).

In an original implementation of electrochemical etching on preprocessed CMOS wafers, Reay et al. [7] fabricated a single crystal n-type silicon well with electronic circuits fully suspended from an oxide support beam. Instead of using KOH, they used TMAH with silicon dissolved in the solution in order to prevent the etch of exposed aluminum bond pads (Figure 3.11).

Plasma-phase etching

Plasma-phase (or dry) etching is a fundamentally important process in the semiconductor industry. Companies such as Applied Materials, Inc., of Santa Clara, California, and Lam Research Corporation of Fremont, California, are leading developers and suppliers of plasma-etching systems of silicon as well as silicon dioxide, silicon nitride, and a wide variety of metals. Conventional plasma-phase etch processes are commonly used

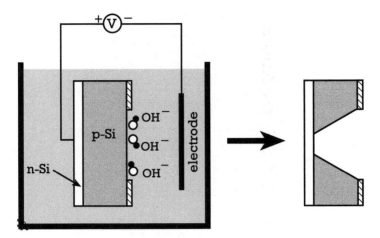

Figure 3.10 Illustration of electrochemical etching using n-type epitaxial silicon. The n-type silicon is biased above its passivation potential so it is not etched. The p-type layer is etched in the solution. The etch stops immediately after the p-type layer is completely removed.

for etching polysilicon in surface micromachining, and for the formation of shallow cavities in bulk micromachining. But the recent introduction of deep reactive ion etching (DRIE) systems by Surface Technology Systems (STS), Ltd., Abercarn, Wales, United Kingdom, PlasmaTherm, Inc., St. Petersburg, Florida, and Alcatel, S.A., Paris, France, provided a new, powerful tool for the etching of very deep trenches (up to 500 μm) with nearly vertical sidewalls.

The basic principle of plasma[3]-phase etching involves the generation of chemically reactive neutrals (e.g., F, Cl) and ions (e.g., SF_x^+) that are accelerated under the effect of an electric or magnetic field towards a target substrate. The reactive species (neutrals and ions) are formed by the collision of molecules in a reactant gas (e.g., SF_6, CF_4, Cl_2, $CClF_3$, and NF_3) with a cloud of energetic electrons excited by an RF field. When the etch process is purely chemical, powered by the spontaneous reaction of neutrals with silicon, it is colloquially referred to as "plasma etching." But if

---

3. Energetic electrons excited in a high-frequency electromagnetic field collide with gas molecules to form ions and reactive neutral species. The mixture of electrons, ions, and neutrals is called a plasma, and constitutes a phase of matter distinct from solids, liquids, or gases. Plasma-phase operation increases the density of ions and neutral species that can participate in a chemical reaction, whether it is deposition or etching, and thus can accelerate the reaction rate.

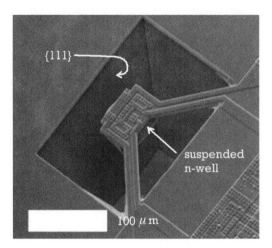

Figure 3.11 A fully suspended *n*-type crystalline silicon island electrochemically etched in TMAH after the completion of the CMOS processing. Courtesy of R. Reay, Linear Technology, Inc., Milpitas, California, and E. Klaassen, IBM, San Jose, California.

ion bombardment of the silicon surface plays a synergistic role in the chemical etch reaction, the process is then referred to as reactive ion etching (RIE). It is the directionality of the accelerated ions that gives RIE its anisotropy. Asymmetric electrodes and low chamber pressures (< 5 Pa) are characteristic of RIE operation. Inductively coupled plasma reactive ion etching (ICP-RIE) provides further excitation to the electron cloud and to the reactive ions by means of an externally applied RF magnetic field. ICP increases the density of ions and neutrals resulting in higher etch rates. The remainder of this section focuses on deep reactive ion etching and its application in micromachining. Further reading on the basics of plasma etching is suggested at the end of this chapter.

Deep reactive ion etching (DRIE) evolved from the need within the micromachining community for an etch process capable of anisotropically etching high aspect ratio trenches at rates substantially larger than 0.1 to 0.5 µm/min, typical of traditional plasma and RIE etchers. In one approach developed by Alcatel, the wafer is cooled to cryogenic temperatures. Condensation of the reactant gases (SF_6 and O_2) protects the sidewalls from etching by the reactive fluorine atoms. However, cryogenic cooling may be difficult to maintain locally, and could result in undesirable thermal stresses. Another approach currently used by

The Toolbox: Processes for Micromachining 67

Alcatel, PlasmaTherm, and STS [12] follows a method patented by Robert Bosch GmbH, of Stuttgart, Germany, in which etch and deposition cycles alternate in an ICP-RIE system [13] (Figure 3.12). The etch cycle, typically lasting 5 to 15 s, uses SF_6 to etch silicon. In the next cycle, a fluorocarbon polymer (made of a chain of CF_2 molecules similar in composition to Teflon™), about 10 nm thick, is plasma deposited using C_4F_8 as a source gas. In the following etch cycle the energetic ions (SF_x^+) remove the protective polymer at the bottom of the trench, but the film remains relatively intact along the sidewalls. The repetitive alternation of the etch and passivation steps results in a very directional etch at rates between 1.5 and 4 μm/min. Some scalloping is observed near the top of the trench, but in general, the sidewalls exhibit good surface planarity with roughness less than 50 nm, which allows their use as optically reflective surfaces. Ongoing research work aims at achieving etch rates above

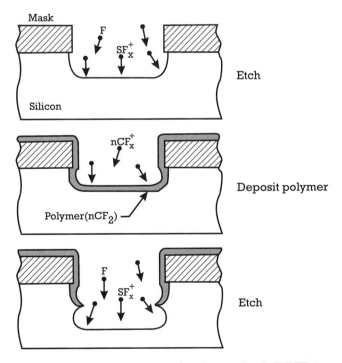

Figure 3.12 Profile of a deep reactive ion etched (DRIE) trench using the Bosch process. The process cycles between an etch step using SF_6 gas, and a polymer deposition step using C_4F_8. The polymer protects the sidewalls from etching by the reactive fluorine neutrals. The scalloping effect of the etch is exaggerated.

5 μm/min by increasing the flow rate of SF_6 and the RF coil excitation power. The latter increases the density of ions available to participate in the etch and deposition cycles, resulting in a net improvement in passivation and etch rates (Table 3.3).

A limitation of DRIE is the dependence of the etch rates on the aspect ratio (ratio of height to width) of the trench. The effect is known as lag or aspect ratio-dependent etching (ARDE) (Figures 3.13 and 3.14). The etch rate is diffusion limited and drops significantly for narrow trenches. A quick remedy is implemented at the mask layout stage by eliminating large disparities in trench widths. The effect of lag can also be greatly alleviated by adjusting the process parameters so that a balance is reached between the diffusion-limited rates of the etch and passivation steps [14]. These parameters are found with experimentation, and may vary depending on the mask layout. The penalty for minimizing lag is typically a reduction in the etch rate to about 1 μm/min.

The high selectivity to silicon dioxide makes it possible to etch deep trenches and stop on a buried layer of silicon dioxide (e.g., silicon-on-insulator wafers). Experiments show that when the etch reaches the buried oxide layer, the concentration of reactive fluorine species increases dramatically, which degrades the passivation layer on the sidewalls. Furthermore, the oxide layer becomes charged. These effects result in an undesirable lateral undercut confined to the silicon-oxide interface. The problem is eliminated by increasing the flow rate of C_4F_8 for a thicker sidewall passivation layer, but at the expense of a decrease in the etch rate to less than 1 μm/min (Figures 3.15(a) and 3.15(b)).

Deep reactive ion etching is a powerful tool for the formation of substantially deep trenches with vertical sidewalls, however process optimization is required for each mask pattern and desired depth.

Figure 3.13 Lag or aspect ratio-dependent etching (ARDE) in DRIE. The etch rate decreases with increasing trench aspect ratio. Courtesy of Lucas NovaSensor, Fremont, California.

The Toolbox: Processes for Micromachining

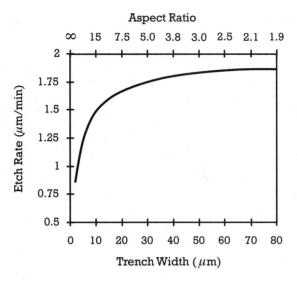

Figure 3.14 Etch rate dependence on feature size and aspect ratio.

Combinations of processes may also be necessary for special situations. For instance, etching deep trenches in silicon-on-insulator (SOI) wafers requires a fast etch process followed by a slow etch process as soon as the buried silicon dioxide layer is detected.

Table 3.3
Process Characteristic of the DRIE Process
Used in the STS and PlasmaTherm Systems

SF_6 flow	30–150 sccm
C_4F_8 flow	20–100 sccm
Etch cycle	5–15 s
Deposition cycle	5–12 s
Pressure	0.25–10 Pa
Temperature	20–80° C
Etch rate	1.5–4 μm/min
Sidewall angle	90° ± 2°
Selectivity to photoresist	~100 to 1
Selectivity to SiO_2	~200 to 1

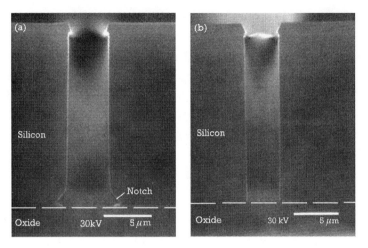

Figure 3.15 (a) Lateral etch observed at the interface between silicon and buried oxide layers. (b) Undercut eliminated with enhanced passivation. Courtesy of Surface Technology Systems, Ltd., Abercarn, United Kingdom.

Advanced process tools

Anodic bonding

Anodic bonding is a simple process that joins together a bare silicon wafer and a sodium-containing glass substrate (e.g., Corning Pyrex® 7740 and 7070, Shott 8330 and 8329). It is fundamental to the manufacture of a wide variety of sensors, including pressure sensors, because it provides a rigid support, in the form of a glass substrate, for the rather fragile silicon wafer.

The bonding is performed at a temperature between 200° and 500° C in a vacuum, air, or an inert gas environment. The application of a large voltage (500–1500 V) across the two substrates, with the glass held at the negative potential, causes mobile positive ions (mostly Na^+) in the glass to migrate away from the silicon-glass interface towards the cathode, leaving behind fixed negative charges (Figure 3.16). The bonding is complete when the ion current (measured externally as an electron current) vanishes, indicating that all mobile ions have reached the cathode. The electrostatic attraction between the fixed negative charges in the glass and positive charges in the silicon holds the two substrates together, and facilitates the chemical bonding of glass to silicon. This is the reason anodic bonding is also known as electrostatic bonding. A very thin

(< 100 nm) silicon dioxide layer on the silicon wafer is sufficient to disturb the current flow and the bond. The coefficient of thermal expansion of the glass substrate is preferably matched to that of silicon in order to minimize thermal stresses. For example, Corning Pyrex® 7740 has a coefficient of thermal expansion of $3.2 \times 10^{-6}/°$ C; silicon's coefficient is $2.6 \times 10^{-6}/°$ C. Sputtered or evaporated glass films containing sodium can be used to anodically bond two silicon substrates. In this case, the required voltage to initiate the bond process decreases to less than 100 V due to the thinness of the glass layer.

Silicon-fusion bonding

Silicon fusion bonding, also known as direct wafer bonding, is a process capable of securely joining two silicon substrates. It emerged as an important step in the development of silicon-on-insulator (SOI) technology during the 1980s for high-frequency and radiation-hardened CMOS applications [15]. SOI wafers made by silicon fusion bonding are commercially available today from many vendors. The concept was quickly extended to the manufacture of pressure sensors [16] and accelerometers in the late 1980's, and is now widely accepted as an important technique in the MEMS toolbox.

The bonding can be between two bare silicon surfaces, or with an intermediate silicon dioxide layer (SOI-type). The bonding mechanism is not well understood, but it is widely believed that it occurs at the

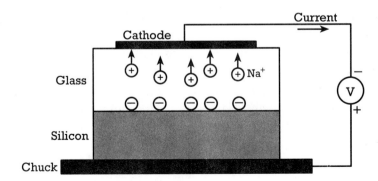

Figure 3.16 Illustration of anodic bonding between glass and silicon. Mobile sodium ions in the glass migrate to the cathode leaving behind fixed negative charges. A large electric field at the silicon glass interface holds the two substrates together, and facilitates the chemical bonding of glass to silicon.

molecular level between silicon and oxygen atoms at the interface. Both wafers are first cleaned in sulfuric acid, followed by hydrochloric acid to remove organic and metal contaminants. Surface cleanliness is necessary to ensure a uniform and void-free bond. The two wafers are then immersed in an ammonium hydroxide solution at approximately 100° C. This "hydration" step serves to provide hydroxyl (OH) groups on the bond surfaces to make them hydrophilic [17]. The bond surfaces are then carefully brought into contact and held together by van der Waals forces. Poor bonding and separation occur when using bowed or nonplanar wafers. A temperature anneal at 800° to 1100° C promotes and strengthens the bond according to the reaction:

$$Si\text{-}O\text{-}H \cdots H\text{-}O\text{-}Si \rightarrow Si\text{-}O\text{-}Si + H_2O.$$

A thin polysilicon film can be fusion-bonded to a silicon wafer or to a silicon dioxide layer if it exhibits a very smooth and planar surface. This can be achieved by using chemomechanical polishing, described in the following section.

In some cases, geometrical features on the two bond surfaces must be aligned to each other prior to bonding. For instance, a cavity in one wafer may be joined to an access port provided through the second wafer. Special equipment is necessary to perform the alignment and bonding. Karl Süss and Electronic Visions, two major equipment manufacturers, use similar schemes to align and bond. The two wafers are typically mounted in a special fixture with the two bond surfaces facing each other, and then aligned in a manner similar to double-sided alignment in lithography. The two bond surfaces are subsequently brought in close proximity to each other, separated by precisely machined thin metal spacers. A mechanical clamping mechanism holds the aligned wafers in position. The spacers are carefully removed bringing the two wafers into contact without loss of alignment. Under proper conditions, the registration (relative misalignment) error can be less than 5 μm. Wafer bonding can be arbitrarily repeated to form thick crystalline multiwafer stacks.

Grinding, polishing, and chemomechanical polishing (CMP)

Some applications require the bonding of thin silicon substrates (< 200 μm) on standard thickness wafers (400 μm for double-sided, polished,

The Toolbox: Processes for Micromachining

100-mm diameter wafers). But thin wafers are very fragile and difficult to handle. Instead, two standard thickness wafers are fusion bonded, and one side is thinned down to the desired thickness. The thickness reduction is achieved using "grinding" and "polishing." The wafer is first mounted on a rotating table and ground by a wheel spinning in the opposite direction with a diamond-based paste. The grinding mechanically abrades silicon and reduces the thickness of the wafer. The resulting surface roughness is removed in the subsequent polishing step in which wafers are mounted inside precise templates on a rotating table. A wheel with a synthetic, felt-like texture polishes the wafer surface using a slurry containing fine diamond particles in a very dilute alkaline solution. The final surface finish is smooth with an overall thickness control of approximately ±5 μm. Damage to the crystal incurred during the grinding step must be annealed at a high temperature (> 1000° C), otherwise defects are preferentially attacked in anisotropic etch solutions.

Chemomechanical polishing (CMP) is a method commonly used in the semiconductor industry for the planarization of interlayer dielectric insulators. The polishing combines mechanical action with chemical etching using an abrasive slurry dispersed in an alkaline solution (pH>10). The rate of material removal is controlled by the slurry flow and pH, applied pressure on the polishing head, rotational speed, and operating temperature. CMP is an excellent planarization method yielding a surface roughness less than 1 nm over large dimensions, but it is slow, with removal rates of less than 100 nm/min, compared to 1 μm/min for standard polishing (Figure 3.17).

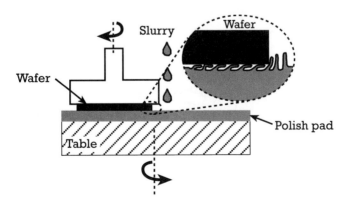

Figure 3.17 Illustration of chemomechanical polishing.

Sol-gel deposition methods

A *sol-gel* process is a chemical reaction between solid particles in colloidal suspension within a fluid (a sol) to form a gelatinous network (a gel) that can be transformed to solid phase upon removal of the solvent. Sol-gel is not a unique process, but rather represents a broad type of processing capable of forming glasses and ceramics in a multitude of shapes, starting from basic chemical precursors. A widespread application of sol-gel processing is in the coating of surfaces with optical absorption or index-graded antireflective materials. It has been used in research laboratories to deposit thick piezoelectric films on silicon substrates (Figure 3.18).

A sol-gel process starts by dissolving appropriate chemical precursors in a liquid to form a sol. Taking the sol through its gel-point transforms it to a gel. This is the point in the phase diagram where the sol undergoes polymerization, and abruptly changes from a viscous liquid state to a gelatinous network. Both sol formation and gelation are low-temperature steps. The gel is then formed into a solid shape (e.g., fiber or lens), or applied as a film coating on a substrate by spinning, dipping, or spraying. For example, tetraethoxysilane (TEOS) in water can be converted into a silica gel by hydrolysis and condensation using hydrochloric acid as a catalyst. Drying and sintering at an elevated temperature (200–600° C) results in the transition of the gel to glass, and then densification to silicon dioxide [18]. Silicon nitride, alumina, and piezoelectric lead-zirconium-titanate (PZT) can also be deposited by sol-gel methods.

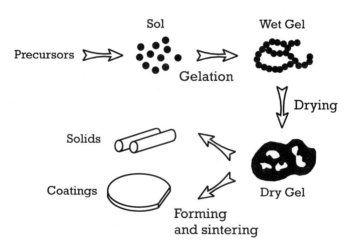

Figure 3.18   Basic flow of a sol-gel process.

The Toolbox: Processes for Micromachining

Electroplating and molding

Electroplating is a well-established industrial method that was adapted in micromachining technology to the deposition of metal films. A variety of metals including gold, copper, nickel, and nickel-iron (Permalloy™) have been electroplated on silicon substrates coated with a suitable thin metal plating base (Table 3.4).

Electroplated structures can be made to take the shape of a mold. The simplest approach to mold preparation is to expose and develop a pattern in thick (10 to 100 μm) resist using optical lithography. The largest aspect ratio achievable is approximately 3, limited by resolution and depth of focus. LIGA replaces optical lithography with x-ray lithography to define very high aspect ratio features (> 100) in very thick (up to 1000 μm) polymethylmethacrylate (PMMA—a pure form of Plexiglas™). Precision reduction gears and other microstructures are common with LIGA, but the method is considered expensive because of the requirement to use collimated x-ray irradiation available only from synchrotrons. Mold formation using optical lithography is often called "poor man's LIGA." Guckel [19] provides additional details on the molding of high aspect ratio structures fabricated with x-ray lithography (Figure 3.19).

Combining the tools—examples of commercial processes

The sequence in which various tools from the toolbox are combined determines a unique microfabrication process. It may be specific to a

Table 3.4
Examples of Solutions for Electroplating Selected Metals

Metal	Solution
Gold	$KAu(CN)_2:K_3C_6H_5O_7:HK_2O_4P:H_2O$
Copper	$CuSO_4:H_2SO_4:H_2O$
Nickel	$NiSO_4:NiCl_2:H_3BO_3:H_2O$
Permalloy™	$NiSO_4:NiCl_2:FeSO_4:H_3BO_3:C_7H_4NNaO_3S:H_2SO_4:H_2O$
Platinum	$H_2PtCl_6:Pb(CH_2COOH)_2:H_2O$
Aluminum	$LiAlH_4:AlCl_3$ in diethyl ether

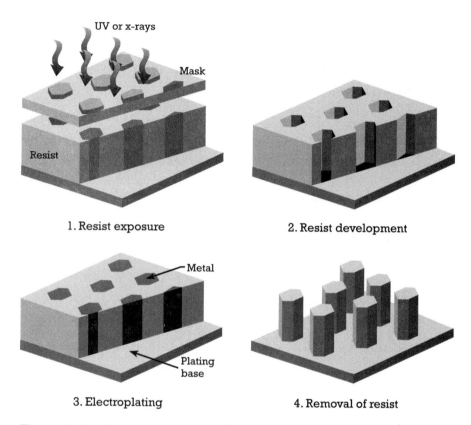

Figure 3.19 Illustration of mold formation using either optical or x-ray lithography and electroplating (LIGA).

particular design, or may be sufficiently general that it can be used to fabricate a broad range of different designs. This section describes three examples of fabrication processes that are generic in their nature, and used today in manufacturing at a number of companies and commercial foundries.

In the first process, thin plates of polysilicon form planar structural elements. This process belongs to a category known as "surface micromachining" where the thickness of the microstructures is very small compared to the overall thickness of the wafer. Originating at the University of California at Berkeley, it is an established manufacturing process at Analog Devices, Inc., Norwood, Massachusetts; Cronos Integrated Microsystems, Inc., Research Triangle Park, North Carolina; and Robert Bosch GmbH, Stuttgart, Germany. The second process combines

The Toolbox: Processes for Micromachining 77

silicon fusion bonding and deep reactive ion etching to form deep microstructures in crystalline silicon. It is now a manufacturing platform at Lucas NovaSensor, Fremont, California. The third process uses isotropic dry etching to release crystalline microstructures formed using a deep etch step. The process, known by its SCREAM (Single Crystal Reactive Etching And Metalization) acronym, was initially developed at Cornell University. EG&G IC Sensors, Milpitas, California, uses a variation on SCREAM for the manufacture of accelerometers. All of these processes are compatible with CMOS fabrication, and hence allow the integration of electronic circuits with microelectromechanical devices. As we will observe next, a key distinguishing feature among them is the "release" step that frees the microstructures in selected locations from the underlying substrate.

Polysilicon surface micromachining

Surface micromachining builds a stack of polysilicon thin films with alternating layers of sacrificial silicon oxide. A typical stack contains a total of four or five layers, but may be more complex. For example, the process at Sandia National Laboratories stacks up to five polysilicon and five oxide layers. The polysilicon films form the structural elements, and are normally deposited using LPCVD followed by a high temperature anneal (> 900°C) for relieving mechanical stresses. The silicon dioxide layer is deposited using CVD. The polysilicon and silicon dioxide layers are each 2-μm-thick; however, Robert Bosch uses a process with 10-μm-thick polysilicon grown by epitaxy over silicon dioxide. Each of the layers in the stack is lithographically patterned and etched before the next layer is deposited in order to form the appropriate shapes, and to make provisions for anchor points to the substrate. The final "release" step consists of etching the silicon dioxide (hence the sacrificial term) to free the polysilicon plates and beams, thus allowing motion in the plane of and perpendicular to the substrate (Figure 3.20). Gears, micromotors, beams, simple as well as hinged plates, and a number of other structures have been demonstrated, though only accelerometers and yaw-rate sensors are currently in volume production. In a similar process, Texas Instruments' Digital Mirror Device™ display technology uses a surface micromachined device with aluminum as its structural element and an organic polymer as a sacrificial layer. Chapter 4 will describe this particular device in greater detail. Surface micromachining offers significant flexibility to fabricate planar structures one layer at a time, but their thinness limits the applications to those benefiting from essentially two-dimensional forms.

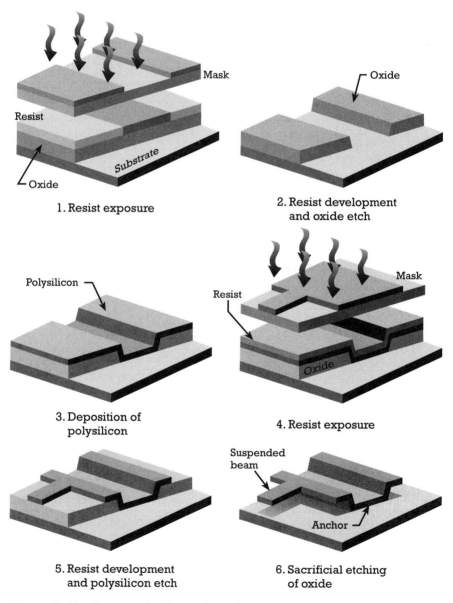

Figure 3.20 Schematic illustration of the basic process steps in surface micromachining.

The release step for the polysilicon-based surface micromachining process consists of immersing the wafers in diluted or buffered solutions of hydrofluoric acid, which laterally etches the silicon dioxide between the polysilicon plates with very high selectivity to silicon. The challenge in a manufacturing environment is to ensure during the subsequent drying of the wafers that the moveable plates and beams do not stick due to capillary action. One solution is "supercritical drying" in carbon dioxide to eliminate capillary forces. This technique transforms the liquid into a supercritical fluid, a state in the phase diagram where the liquid and gas phases are indistinguishable, then to a gas that is gently vented. Another way to avoid sticking is to coat the mechanical microstructures with a hydrophobic passivation layer, such as a Teflon™-like fluorocarbon polymer. Bustillo et al. present a comprehensive review of surface micromachining in a special issue of the Proceedings of the IEEE on MEMS [20].

Combining silicon fusion bonding with reactive ion etching (SFB-DRIE)

The SFB-DRIE process involves the formation of tall structures in crystalline silicon to overcome the thinness limitation of surface micromachining [21]. Instead of depositing thin polysilicon layers, crystalline silicon substrates are fusion-bonded to each other in a stack. Each substrate is polished down to a desired thickness, then patterned and etched before the next one is bonded. The intermediate silicon dioxide between the silicon substrates is not a sacrificial layer, but is rather for electrical and thermal insulation. The process allows the building of complex three-dimensional structures one thick layer at a time.

The basic process flow begins by etching a cavity in a first wafer—the handle wafer. A second wafer is silicon fusion-bonded. An optional polishing step can reduce the thickness of the bonded wafer to any desired value. CMOS electronic circuits can then be integrated on the top surface of the bonded stack without affecting any of its mechanical properties. Finally, a deep reactive ion etching step (DRIE) delineates the geometrical shape of the microstructures, and mechanically releases them as soon as the etch reaches the embedded cavity. This cavity takes the role of the sacrificial layer in surface micromachining, and ensures that the micromechanical structures are free to move, except at well-defined anchor points (Figure 3.21).

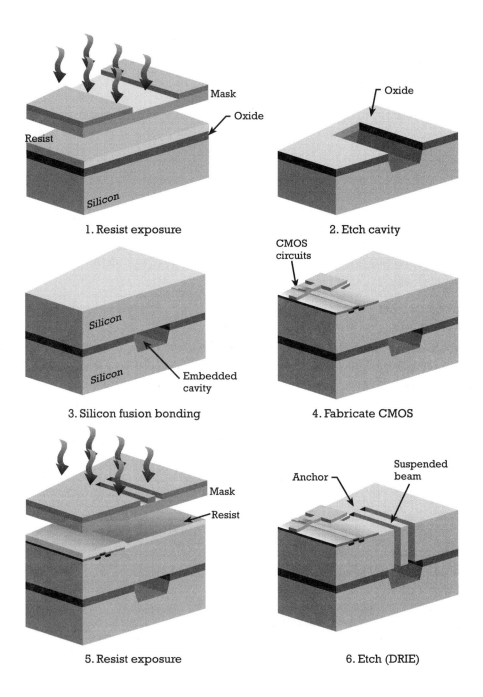

Figure 3.21 Fabrication process combining silicon fusion bonding and deep reactive ion etching.

The Toolbox: Processes for Micromachining

The high aspect ratio and depth available using the SFB-DRIE process add new dimensions to the design and fabrication of complex three-dimensional structures. A range of new applications, including those integrating fluid-flow functions such as valving and pumping can be addressed with this process. Robust actuators made of crystalline silicon are also feasible with an available output force approaching one newton (Figure 3.22).

SCREAM

The SCREAM process [22] uses yet another approach to release crystalline microstructures. Standard lithography and etching methods define trenches between 10 and 50 μm in depth, which are then protected with a conformal layer of PECVD silicon oxide. An anisotropic etch step selectively removes the protective oxide only at the bottom of the trench. A subsequent plasma silicon etch extends the depth of the trench. A dry isotropic etch step using sulfur hexafluoride (SF_6) laterally etches the exposed sidewalls near the bottom of the trench, thus undercutting

Figure 3.22 Photograph of a 200-μm-deep thermal actuator fabricated using silicon fusion bonding and deep reactive ion etching. Courtesy of Lucas NovaSensor, Fremont, California.

adjacent structures and mechanically releasing them. Sputter deposition of aluminum provides the metal for electrical contacts and interconnects (Figure 3.23).

Summary

The toolbox of micromachining processes is very large and diverse. The vast majority of the methods can be condensed into three major categories:

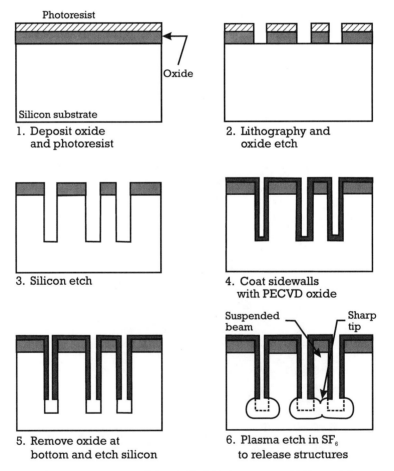

Figure 3.23 Basic steps of the SCREAM process. Adapted from Shaw et al. [22].

The Toolbox: Processes for Micromachining 83

- Material deposition, including thin film deposition and bonding processes;
- Pattern definition using lithography;
- Etching.

A complete micromachining process flow consists of a series of steps using a number of methods from the toolbox to build complex microstructures one layer at a time.

References

[1] Katz, L. E., "Oxidation." In *VLSI Technology*, pp. 131–167, S. M. Sze (ed.), New York, NY: McGraw-Hill, 1983.

[2] Thornton, J. A., and D. W. Hoffman, "Stress Related Effects in Thin Films," *Thin Solid Films*, Vol. 171, 1989, pp. 5–31.

[3] Williams, K. R., and R. S. Muller, "Etch Rates for Micromachining Processing," *Journal of Microelectromechanical Systems*, Vol. 5, No. 4, Dec. 1996, pp. 256–269.

[4] Seidel, H., et al., "Anisotropic Etching of Crystalline Silicon in Alkaline Solutions," *Journal of Electrochemical Society*, Vol. 137, No. 11, Nov. 1990, pp. 3612–3632.

[5] Kovacs, G. T. A., N. I. Maluf., and K. E. Petersen, "Bulk Micromachining of Silicon," in *Integrated Sensors, Microactuators, & Microsystems (MEMS)*, pp. 1536–1551, K. D. Wise (ed.), Proceedings of the IEEE, Vol. 86, No. 8, Aug. 1998.

[6] Schnakenberg, U., W. Benecke, and P. Lange, "TMAHW Etchants for Silicon Micromachining," *Proc. 1991 Int. Conf. on Solid-State Sensors and Actuators*, San Francisco, CA, June 24–27, 1991, pp. 815–818.

[7] Reay, R. J.,E. H. Klaassen, and G. T. A. Kovacs, "Thermally and Electrically Isolated Single-Crystal Silicon Structures in CMOS Technology," *IEEE Electron Device Letters*, Vol. 15, Oct. 1994, pp.309–401.

[8] Ammar, E. S., and T. J. Rodgers, "UMOS Transistors on (110) Silicon," *IEEE Transactions on Electron Devices*, Vol. ED-27, No. 5, May 1980, pp. 907–914.

[9] Sandmaier, H., et al., "Compensation Techniques in Anisotropic Etching of (100)-Silicon Using Aqueous KOH," *Proc. 1991 Int. Conf. on Solid-State Sensors and Actuators*, San Francisco, CA, June 24–27, 1991, pp. 456–459.

[10] Waggener, H. A., "Electrochemically Controlled Thinning of Silicon," *Bell System Technology Journal*, Vol. 50, 1970, pp. 473–475.

[11] Kloeck, B, et al., "Study of Electrochemical Etch-Stop for High Precision Thickness Control of Silicon Membranes," *IEEE Transactions on Electron Devices*, Vol. 36, No. 4., Apr. 1989, pp. 663–669.

[12] Bhardwaj, J., and H. Ashraf, "Advanced Silicon Etching using High Density Plasmas," *Proc. SPIE, Micromachining and Microfabrication Process Technology Symp.*, Austin, TX, Oct. 23–24, 1995, Vol. 2639, pp. 224–233.

[13] Lärmer, F., and P. Schilp, "Method of Anisotropically Etching Silicon," German Patent DE 4 241 045, 1994.

[14] Ayón, A. A., et al., "Etching Characteristics and Profile Control in a Time Multiplexed Inductively Coupled Plasma Etcher," *Tech. Digest Solid-State Sensor and Actuator Workshop*, Hilton Head Island, SC, June 8–11, 1998, pp. 41–44.

[15] Lasky, J. B., "Wafer Bonding for Silicon-On-Insulator Technologies," *Applied Physics Letters*, Vol. 48, No. 1, Jan. 6, 1986, pp. 78–80.

[16] Petersen, K. E., et al., "Silicon Fusion Bonding for Pressure Sensors," *Tech. Digest Solid-State Sensor and Actuator Workshop*, Hilton Head Island, SC, June 6–9, 1988, pp. 144–147.

[17] Tong, Q. -Y., and U. Gösele, *Semiconductor Wafer Bonding*, New York, NY: Wiley, 1999, pp. 49–72.

[18] Strawbridge, I., and P. F. James, "Glass Formation from Gels." In *High Performance Glasses*, pp. 20–49, M. Cable and J. M. Parker (eds.), London, England: Blackie Publishing, 1992.

[19] Guckel, H., "High-Aspect Ratio Micromachining Via Deep X-Ray Lithography," in *Integrated Sensors, Microactuators, & Microsystems (MEMS)*, pp. 1586–1593, K. D. Wise (ed.), Proceedings of the IEEE, Vol. 86, No. 8, Aug. 1998.

[20] Bustillo, J. M., R. T. Howe, and R. S. Muller, "Surface Micromachining for Microelectromechanical Systems," in *Integrated Sensors, Microactuators, & Microsystems (MEMS)*, pp. 1559–1561, K. D. Wise (ed.), Proceedings of the IEEE, Vol. 86, No. 8, Aug. 1998.

[21] Klaassen, E. H., et al., "Silicon Fusion Bonding and Deep Reactive Ion Etching; A New Technology for Microstructures," *Proc. 8th Int. Conf. on Solid-State Sensors and Actuators*, Stockholm, Sweden, June 25–29, 1995, pp. 556–559.

[22] Shaw, K. A., Z. L. Zhang, and N. C. MacDonald,, "SCREAM-I: A Single Mask, Single-Crystal Silicon, Reactive Ion Etching Process for Microelectromechanical Structures," *Sensors and Actuators*, Vol. A40, No. 1, 1994, pp. 63–70.

Selected bibliography on VLSI microfabrication

Flamm, D. L., and G. K. Herb., "Plasma Etching Technology." In *Plasma Etching: An Introduction*, pp. 1–89, D. M. Manos and D. L. Flamm (eds.), San Diego, CA: Academic Press, 1989.

Kamins, T., *Polycrystalline Silicon for Integrated Circuits*, Boston, MA: Kluwer Academic Publishers, 1988.

Moreau, M., *Semiconductor Lithography Principles*, Practices and Materials, New York, NY: Plenum Press, 1988.

ULSI Technology, C. Y. Chang and S. M. Sze (eds.), New York, NY: McGraw-Hill, 1996.

Selected bibliography on micromachining

Kovacs, G. T. A., *Micromachined Transducers Sourcebook*, New York, NY: McGraw-Hill, 1998.

Madou, M., *Fundamentals of Microfabrication*, Boca Raton, FL: CRC Press, 1997.

Tong, Q. -Y., and U. Gösele, *Semiconductor Wafer Bonding*, New York, NY: Wiley, 1999.

Wise, K. D., Ed., "Special Issue on Integrated Sensors, Microactuators, and Microsystems (MEMS)," *Proceedings of the IEEE*, Vol. 86, No. 8, August 1998.

CHAPTER 4

The Gearbox: Commercial MEM Structures and Systems

Contents

General design methodology
Techniques for sensing and actuation
Passive MEM structures
Sensors
Actuators
Summary

> ... for I believe that his device had tremendous advantages and unless there be other systems of equal merits which are unknown to me, I am of the opinion that he has the most remarkable system in existence.
>
> *David Sarnoff on E. Howard Armstrong's radio receiver, 1914. Quoted in the "Empire of the Air," by Tom Lewis.*

Armed with an understanding of the fabrication methods, it is time to examine various types of microelectromechanical (MEM) structures and systems. It is apparent that with a vast and diverse set of fabrication tools, creativity abounds. Indeed, the list of MEM structures and devices continues to grow daily as more applications prove to benefit from miniaturization. But just as necessity is the mother of all inventions, it is economics that ultimately determines the commercial success of a particular design or technology. Demonstrations of micromachined devices are innumerable, but the successful products are few. MEMS technology is only a means to achieve a solution for a particular application. A quest for its perfection should not entail an oversight of the end objective: The application itself.

This chapter begins with a short introduction on the general methodology of the design process and a listing of commonly used sensing and actuation techniques, followed by a review of silicon MEM structures and systems that exist as off-the-shelf commercial products, or have been incorporated into commercially available systems.

Three general categories form the total extent of MEMS: sensors, actuators, and passive structures. Sensors are transducers that convert mechanical, thermal, or other forms of energy into electrical energy; actuators do exactly the opposite. Passive structures are devices in which no transducing occurs. A complete listing of all MEMS demonstrations is not sought here; rather the theme is to illustrate the state of the technology by providing sufficient examples of structures and systems that have proven their commercial viability, or show promise to do so in the near future.

General design methodology

Starting with a list of specifications for the MEM device or system, the design process begins with the identification of the general operating principles and overall structural elements, then proceeds onto analysis and simulation, and finally onto outlining of the individual steps in the fabrication process. This is often an iterative process involving continuous adjustments to the shape, structure, and fabrication steps. The layout of the lithographic masks is the final step before fabrication, and is completed using specialized computer-aided design (CAD) tools to define the two-dimensional patterns.

Early design considerations include the identification of the general sensing or actuation mechanisms based on performance requirements. For instance, output force requirement of a mechanical microactuator may favor thermal or piezoelectric methods and preclude electrostatic ones. Similarly, the choice of piezoresistive sensing is significantly different from capacitive or piezoelectric sensing. The interdisciplinary nature of the field brings together considerations from a broad range of

The Gearbox: Commercial MEM Structures and Systems

specialties including mechanics, optics, fluid dynamics, materials science, electronics, chemistry, and even biological sciences. On occasion, determining a particular approach may rely on economic considerations or ease of manufacture rather than performance. For example, the vast majority of pressure sensors use cost-effective piezoresistive sense elements instead of the better performing, but more expensive, resonant-type sense structures.

The design process is not an exact analytical science, but rather involves developing engineering models, many for the purpose of obtaining basic physical insights. Computer-based simulation tools that use finite element modeling are convenient to analyze complex systems. A number of available programs, such as ANSYS® (ANSYS, Inc., Canonsburg, Pennsylvania), and MEMCAD (Microcosm Technologies, Inc., Research Triangle, North Carolina) can simulate mechanical, thermal, and electrostatic structures (Figure 4.1). Substantial efforts are currently underway to develop sophisticated programs that can handle coupled multi-mode problems, for example, simultaneously combining fluid dynamics with thermal and mechanical analysis. As powerful as these tools are perceived to be, their universal predictive utility is questionable, but they can provide valuable insight and visualization of the device's operation.

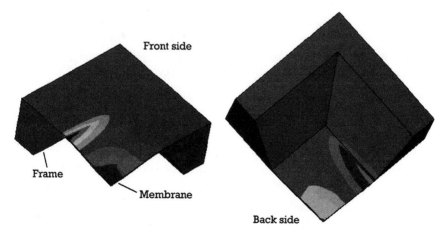

Figure 4.1 A finite element simulation using ANSYS® modeling program of a bulk micromachined silicon pressure sensor showing contours of mechanical stress in response to an applied pressure load.

In planning a fabrication process, the choice is either to use a standard foundry service, or to design a custom process specific to the device or system. If the production unit volume is not sufficiently large, it may be challenging to identify reputable manufacturing facilities willing to develop and implement custom processes.

Techniques for sensing and actuation

Common sensing methods

Sensing is by no means a modern invention. There are numerous historical accounts describing the measurement of physical parameters—most notably, distance, weight, time, and temperature. Early Chinese attempts at making compasses date back to the 12th century with the use of lodestone, a naturally occurring magnetic ore. Modern sensing methods derive their utility from the wealth of scientific knowledge accumulated over the past two centuries. We owe our intimate familiarity with electrostatics and capacitance to the work of Charles Augustin de Coulomb of France and John Priestly of England in the late 18th century; and recall that Lord Kelvin's discovery of piezoresistivity in 1856 is recent in historical terms. What distinguishes these modern techniques is the ability to sense with greater accuracy and stability; and what makes them suitable for microelectromechanical systems is their scalable functionality.

The objective of modern sensing is the transducing of a specific physical parameter, to the exclusion of other interfering parameters, into electrical energy. Occasionally, an intermediate conversion step takes place. For example, pressure or acceleration is converted into mechanical stress, which is then converted to electricity. Perhaps the most common of all modern sensing techniques is temperature measurement using the dependence of various material properties on temperature. This effect is pronounced in the electrical resistance of metals—the temperature coefficient of resistance (TCR) of most metals ranges between 10 and 100 parts per million per degree centigrade.

Piezoresistivity and piezoelectricity are two sensing techniques described in greater detail in Chapter 2. Impurity-doped silicon exhibits a piezoresistive behavior which lies at the core of many pressure and acceleration sensor designs. Measuring the change in resistance and

The Gearbox: Commercial MEM Structures and Systems

amplifying the corresponding output signal tend to be rather simple, requiring a basic knowledge of analog circuit design. A drawback of silicon piezoresistivity is its strong dependence on temperature that must be compensated for with external electronics.

In contrast, capacitive sensing relies on an external physical parameter changing either the spacing or the relative dielectric constant between the two plates of a capacitor. For instance, an applied acceleration pushes one plate closer to the other. Or in the example of relative humidity sensors, the dielectric is an organic material whose permittivity is a function of moisture content. The advantages of capacitive sensing are very low power consumption and relative stability with temperature. Additionally, the approach offers the possibility of electrostatic actuation to perform closed-loop feedback. The following section on actuation methods explains this point further. Naturally, capacitive sensing requires external electronics to convert changes in capacitance into an output voltage. Unlike measuring resistance, these circuits can be substantially intricate if the change in capacitance is small. This is frequently the case in MEMS where capacitance values are on the order of 1 pF and less.

Yet another sensing approach utilizes electromagnetic signals to detect and measure a physical parameter. Magnetoresistive sensors on the read heads of high-density computer disk drives measure the change in conductivity of a material slab in response to the magnetic field of the storage bit. In Hall effect devices, a magnetic field induces a voltage in a direction orthogonal to current flow. Hall effect sensors are extremely inexpensive to manufacture, and make excellent candidates to measure wheel velocity in vehicles. Another form of electromagnetic transducing uses Faraday's law to detect the motion of a current-carrying conductor through a magnetic field. Two yaw-rate sensors described later in this chapter make use of this phenomenon. The control electronics for magnetic sensors can be readily implemented using modern CMOS technology. But generating magnetic fields often necessitates the presence of a permanent magnet or a solenoid (Table 4.1).

Common actuation methods

A complete shift in paradigm becomes necessary to think of actuation on a miniature scale—a four-stroke engine is not scalable. The next five schemes illustrate the diversity and the myriad of actuation options

Table 4.1
The Relative Merits of Piezoresistive, Capacitive, and Electromagnetic Sensing Methods

Piezoresistive	Capacitive	Electromagnetic
Simple fabrication	Simple mechanical structure	Structural complexity varies
Low cost	Low cost	Complex packaging
Voltage or current drive	Voltage drive	Current drive
No need for circuits	Requires electronic circuits	Simple control circuits
Low output impedance	Susceptible to EMI	Susceptible to EMI
High temperature dependence	Low temperature dependence	Low temperature dependence
Small sensitivity	Large dynamic range	Sensitivity ∝ magnetic field
Insensitive to parasitic resistance	Sensitive to parasitic capacitance	Insensitive to parasitic inductance
Open loop	Open or closed loop	Open or closed loop
Medium power consumption	Low power consumption	Medium power consumption

available in MEMS. They are *electrostatic, piezoelectric, thermal, magnetic,* and *phase recovery using shape-memory alloys*. The choice of actuation depends on the nature of the application, ease of integration with the fabrication process, and economic justification. Examples of each actuation method will arise throughout this chapter and the next (Table 4.2).

Electrostatic actuation

Electrostatic actuation relies on the attractive force between two plates or elements carrying opposite charges. A moment of thought quickly reveals that the charges on two objects with an *externally applied* potential between them can only be of opposite polarities. Therefore, an applied voltage, regardless of its polarity, always results in an attractive electrostatic force. If C is the capacitance between two parallel plates, x is the spacing between them, and V is an externally applied voltage, the electrostatic force is then $\frac{1}{2}CV^2/x$. For a spacing of one micrometer, an applied voltage of 5 V, and a reasonable area of $1,000\,\mu m^2$, the electrostatic force is merely 0.11 μN. A natural extension of electrostatic actuation is closed-loop feedback in systems employing capacitive sensing. When sense

The Gearbox: Commercial MEM Structures and Systems

circuits detect the two plates of a capacitor separating under the effect of an external force (e.g., acceleration), an electrostatic feedback voltage is immediately applied by the control electronics to counteract the disturbance and maintain a fixed capacitance. The magnitude of the feedback voltage then becomes a measure of the disturbing force. This feature is integral to the closed-loop operation of many accelerometers and yaw-rate sensors.

Piezoelectric actuation

Piezoelectric actuation can provide significantly large forces, especially if thick piezoelectric films are used. Commercially available piezoceramic cylinders can provide up to a few newtons of force with applied potentials on the order of a few hundred volts. However, thin-film ($< 5\,\mu$m) piezoelectric actuators can only provide a few millinewtons. Both piezoelectric and electrostatic methods offer the advantage of low power consumption since the electric current is very small.

Thermal actuation

Thermal actuation consumes more power than electrostatic or piezoelectric actuation, but can provide, despite its gross inefficiencies, actuation forces on the order of hundreds of millinewtons or higher. At least three distinct approaches have emerged within the MEMS community. The first capitalizes on the difference in the coefficients of thermal expansion between two joined layers of dissimilar materials to cause bending with temperature—the classic case of a bimetallic thermostat studied by S. Timoshenko in 1925 [1]. One layer expands more than the other as temperature increases. This results in stresses at the interface and consequent bending of the stack. The amount of bending depends on the difference in coefficients of thermal expansion and absolute temperature. Unfortunately, the latter dependence severely limits the operating temperature range—otherwise, the device may actuate prematurely on a hot day.

In another approach known as thermopneumatic actuation, a liquid is heated inside a sealed cavity. Pressure from expansion or evaporation exerts a force on the cavity walls, which can bend if made sufficiently compliant. This method also depends on the absolute temperature of the actuator. Valves employing each of the above methods will be described later in this chapter.

A third distinct method utilizes a suspended beam of a same homogeneous material with one end anchored to a supporting frame of the same material [2]. Heating the beam to a temperature above that of the frame causes a differential elongation of the beam's free end, with respect to the frame. Holding this free end stationary gives rise to a force proportional to the beam length and temperature differential. Such an actuator delivers a maximal force with zero displacement, and conversely, no force when the displacement is maximal. Designs operating between these two extremes can provide both force and displacement. A system of mechanical linkages can optimize the output of the actuator by trading off force for displacement, or vice-versa. Actuation in this case is independent of fluctuations in ambient temperature because it relies on the difference in temperature between the beam and the supporting frame.

Magnetic actuation

Lorentz forces form the dominant mechanism in magnetic actuation on a miniaturized scale. This is largely due to the difficulty in depositing permanently magnetized thin films. Electrical current in a conductive element that is located within a magnetic field gives rise to an electromagnetic force—the Lorentz force—in a direction perpendicular to the current and magnetic field. This force is proportional to the current, magnetic field, and length of the element. A conductor 1 mm in length carrying 10 mA in a 1-T magnetic field is subject to a force of 10 μN. Lorentz forces are useful for closed-loop feedback in systems employing electromagnetic sensing. Two yaw-rate sensors described later make use of this method.

Actuation using shape-memory alloys

Finally, of all five schemes, shape-memory alloys undoubtedly offer the highest energy density available for actuation (Table 2.4). The effect, introduced in Chapter 2, can provide very large forces when the temperature of the material rises above the critical temperature, typically around 100° C. The challenge with shape-memory alloys lies in the difficulty of integrating their fabrication with conventional silicon manufacturing processes.

Table 4.2
Comparison of Various Actuation Methods on the
Basis of Maximum Energy Density[1]

Actuation	Max. energy density	Physical & material parameters	Estimated conditions	Approximate order (J/cm3)
Electrostatic	$\frac{1}{2} \varepsilon_0 E^2$	E = electric field ε_0 = dielectric permittivity	5 V/μm	~ 0.1
Thermal	$\frac{1}{2} Y (\alpha \Delta T)^2$	α = coefficient of expansion ΔT = temperature rise Y = Young's modulus	3×10^{-6}/°C 100° C 100 GPa	~ 5
Magnetic	$\frac{1}{2} B^2/\mu_0$	B = magnetic field μ_0 = magnetic permeability	0.1 T	~ 4
Piezoelectric	$\frac{1}{2} Y (d_{33} E)^2$	E = electric field Y = Young's modulus d_{33} = piezoelectric constant	30 V/μm 100 GPa 2×10^{-12} C/N	~ 0.2
Shape-memory alloy	—	Critical temperature		~ 10 [from reports in literature]

[1] Actual energy output may be substantially lower depending on the overall efficiency of the system.

Passive MEM structures

Fluid nozzles

Nozzles are undoubtedly among the simplest microstructures to fabricate using anisotropic etching of silicon, or laser drilling of a metal sheet. A series of United States patents issued in the 1970s to IBM Corporation [3] describe the fabrication of silicon nozzles and their application for ink-jet printing. The Ford Motor Company experimented in the 1980s with silicon nozzles for engine fuel injection. With the expiration of most key patents on nozzle formation, silicon nozzles are becoming common features in the design of atomizers, medical inhalers, and fluid spray systems. For example, SprayChip Systems of Seabrook, Maryland is commercializing a micromachined nozzle using a silicon two-wafer stack capable of accurate control of droplet size. Nozzles need not necessarily be of silicon. MicroParts GmbH, of Dortmund, Germany manufactures a drug-inhaling device for asthma patients that incorporates a precise plastic nozzle, fabricated using the electroplating and molding process described in the previous chapter.

A simple square silicon nozzle can be readily fabricated by depositing silicon nitride on both sides of a {100} wafer and patterning a square in the silicon nitride layer on the back side. Anisotropic etching in potassium hydroxide (KOH) or tetramethyl ammonium hydroxide (TMAH) forms a port through the wafer defined by the {111} planes of silicon. The dimensions of the backside opening in the silicon nitride must be larger than 71% of the wafer thickness in order to etch through the wafer (Figure 4.2).

Forming nozzles of circular or arbitrary shape involves additional fabrication steps. The most common approach is to grow on a {100} wafer a *p*-type epitaxial layer of silicon with a high boron concentration

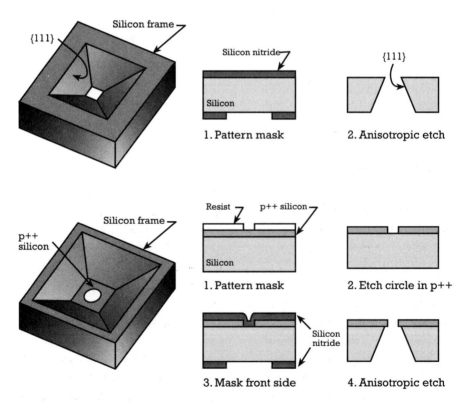

Figure 4.2 Schematic illustrations of square and circular nozzles with their corresponding fabrication steps.

The Gearbox: Commercial MEM Structures and Systems

($> 1 \times 10^{19}$ cm^{-3}). The shape of the nozzle is patterned and etched into the p-type silicon layer using standard lithography and plasma etching (or RIE). A protective layer of silicon nitride is deposited on both sides of the wafer and patterned in the shape of a square on the back side. Double-sided lithography provides accurate alignment between the nozzle opening and the square on the back side. The fabrication is complete with the anisotropic etching of the silicon from the back side using KOH or TMAH. The p-type layer acts as an etch stop, thus preserving the shape of the nozzle.

Choosing a fluid flow path in the plane of the silicon wafer provides further flexibility in shaping the nozzle and the orifices. In an implementation of a CO_2 cleaning apparatus [4], a silicon micromachined nozzle was specially designed to allow subsonic fluid flow at the inlet, and supersonic flow at the outlet. Deep-reactive-ion etching is a suitable process for defining in the silicon a deep channel (50 to 500 μm) following the desired contour of the nozzle. The dimensional control is limited in the plane of the wafer by the lithography to better than one micrometer; whereas in the vertical depth direction, it is limited by the etch process to approximately 10% of the total depth. A top cover is later bonded using anodic bonding of glass or silicon-fusion bonding (Figure 4.3).

Inkjet print nozzles

The thermal inkjet print head, ubiquitous in today's printers for personal computers, receives frequent mention as a premier success story of

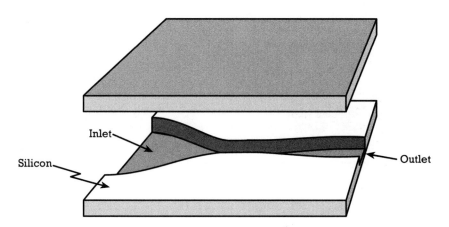

Figure 4.3 Illustration of a nozzle with fluid flow in the plane of the silicon wafer. Adapted from Kneisel et al. [4].

MEMS technology. While thermal inkjet technology is a commercial success for Hewlett-Packard, Inc. of Palo Alto, California and a few other companies, there is little in it that originates from silicon MEMS *per se*. Rather, early inkjet heads used electroplated nickel nozzles. More recent models use nozzle plates drilled by laser ablation. Silicon micromachining is not likely to compete with these traditional technologies on a cost basis. However, applications that require high-resolution printing will probably benefit from micromachined nozzles. At a resolution of 1200 dots per inch (dpi), the spacing between adjacent nozzles in a linear array is a mere 21 µm. Nevertheless, while silicon micromachined inkjet nozzles are not yet commercial, high-performance inkjet technology represents an excellent illustration of how a micromachined structure could potentially become a critical and enabling element in a more complex system.

The device from Hewlett-Packard illustrates the basic principle of thermal inkjet printing [5]. A tantalum thin-film resistor residing on top of electronic control circuits superheats a thin ink layer, beneath an exit nozzle, to 250° C. Within 5 µs, a bubble forms with peak pressures reaching 1.4 MPa (200 psi) and begins to expel ink out of the orifice. After 15 µs, the ink droplet is ejected from the nozzle. Within 24 µs of the firing pulse, the tail of the ink droplet separates and the bubble collapses inside the nozzle, resulting in high cavitation pressures. Within less than 50 µs, the chamber refills and the ink meniscus at the orifice settles (Figure 4.4).

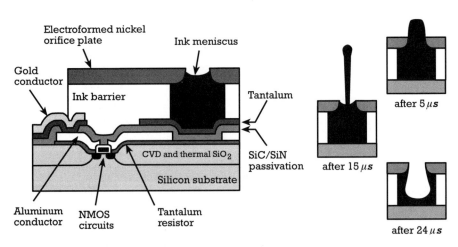

Figure 4.4 Cross-sectional illustration of a Hewlett-Packard thermal inkjet head and the ink firing sequence. Adapted from C. C. Beatty [5].

Sensors

Pressure sensors

The first high volume production of a pressure sensor began in 1974 at National Semiconductor Corporation of Santa Clara, California. Pressure sensing has since grown to a large market, with an estimated 45 million silicon micromachined pressure sensors manufactured in 1998. Nearly all units use bulk micromachining technology. Manifold absolute pressure (MAP) [6] and disposable blood pressure [7] sensing are the two largest applications. The vast majority use piezoresistive sense elements to detect stress in a thin silicon diaphragm in response to a pressure load. A few designs use capacitive methods to sense the displacement of a thin diaphragm.

The basic structure of a piezoresistive pressure sensor consists of four sense elements in a Wheatstone bridge configuration to measure stress within a thin, crystalline silicon membrane (Figure 4.5). The stress is a direct consequence of the membrane deflecting in response to an applied pressure differential across the front and back sides of the sensor. The stress is, to a first order approximation, linearly proportional to the applied pressure differential. The membrane deflection is typically less than one micrometer. The output at full-scale applied pressure is a few millivolts per volt of bridge excitation (the supply voltage to the bridge). The output normalized to input applied pressure is known as sensitivity [(mV/V)/Pa] and is directly related to the piezoresistive coefficients, $\pi_{//}$ and $\pi_\perp$ (see Chapter 2). The thickness and geometrical dimensions of the membrane affect the sensitivity, and consequently, the pressure range of the sensor. Devices rated for low pressure (less than 10 kPa) usually incorporate complex membrane structures, such as central bosses, to improve sensitivity.

A common design layout on {100} substrates positions the four diffused piezoresistive sense elements at the points of highest stress, which occur at the center edges of the diaphragm. Two elements have their primary axes parallel to the membrane edge, resulting in a decrease in resistance with membrane bending. The other two resistors have their axes perpendicular to the edge, which causes the resistance to increase with the pressure load. Other layouts are also possible including designs to measure shear stress, but the main objective remains to position the resistors in the areas of highest stress concentration, in order to maximize

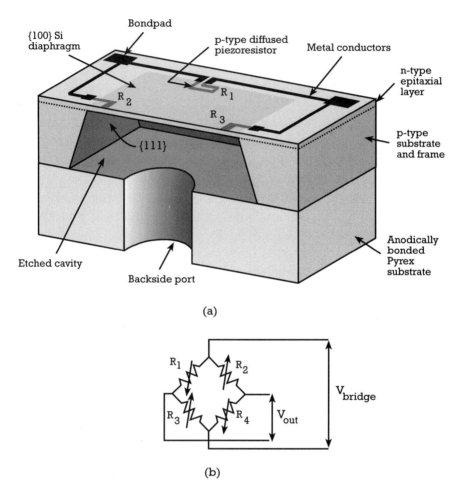

Figure 4.5 (a) Schematic illustration of a pressure sensor with diffused piezoresistive sense elements. (b) The four sense elements form a Wheatstone bridge configuration.

the response to applied pressure. It is necessary that the four piezoresistors have identical resistances in the absence of applied pressure. Any mismatch in resistance, even one caused by temperature, causes an imbalance in the Wheatstone bridge. The resulting output reading is known as zero offset, and is undesirable.

Deep diffusions degrade the sensitivity of the piezoresistors by averaging the stress over the depth of the sense element. Shallow diffusions are prone to surface-charge effects that can cause long-term drift in the output signal. Remedies to these conflicting requirements are frequently

The Gearbox: Commercial MEM Structures and Systems

proprietary to the manufacturers. U.S. patent #4,125,820 (Nov. 14, 1978) assigned to Honeywell, Inc., Minneapolis, Minnesota, illustrates one solution in which the piezoresistive diffusions are buried below the surface of the membrane.

The fabrication process of a typical pressure sensor relies mostly on steps standard to the integrated circuit industry, with the exception of the precise forming of the thin membrane using electrochemical etching (ECE) (Figure 4.6). An *n*-type epitaxial layer of silicon is grown on a *p*-type {100} wafer. A thin, preferably stress-free, insulating layer is deposited or grown on the front side of the wafer, and a protective silicon nitride film is deposited on the back side. The piezoresistive sense

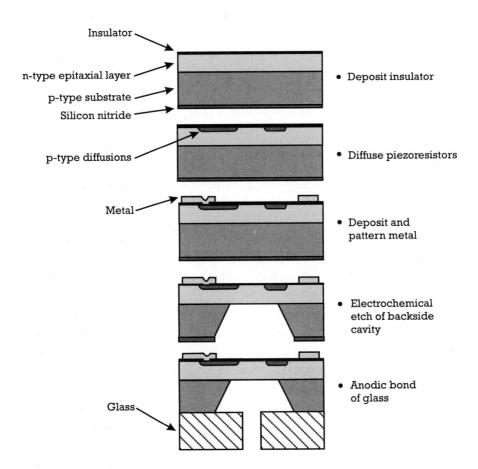

Figure 4.6 Fabrication steps for a piezoresistive gauge, or differential, bulk micromachined pressure sensor.

elements are formed by locally doping the silicon p-type using the masked implantation of boron, followed by a high-temperature diffusion cycle. Etching of the insulator on the front side provides contact openings to the underlying piezoresistors. A metal layer, typically aluminum, is then sputter-deposited and patterned in the shape of electrical conductors and bond pads. A square opening is patterned and etched in the silicon nitride layer on the back side. Double-sided lithography ensures that the backside square is precisely aligned to the sense elements on the front side. At this point, electrical contacts are made to the p-type substrate and n-type epitaxial layer, and the silicon is electrochemically etched from the back side in a solution of potassium hydroxide. Naturally, the front side must be protected during the etch. The etch stops as soon as the p-type silicon is completely removed, and the n-type layer is exposed. The process forms a membrane with precise thickness defined by the epitaxial layer. Anodic bonding in vacuum of a Pyrex® glass wafer on the back side produces an absolute pressure sensor, which measures the pressure on the front side in reference to the cavity pressure (often, vacuum). For differential- or gauge-type pressure sensors, precisely drilled holes in the glass wafer provide vent ports.

The advent of silicon-fusion bonding in the 1980s proved very useful to the design of bulk micromachined pressure sensors. The outward sloping of {111} planes delineating the sensor's frame results in an unnecessary increase in die size. Silicon-fusion bonding allows the forming of the membrane, after the etching of a reference cavity with inward sloping {111} walls. Consequently, extremely small pressure sensors are feasible. For example, Lucas NovaSensor, Fremont, California, manufactures a sensor that is 400-μm-wide, 800-μm-long, and 150-μm-thick, and fits inside the tip of a catheter (Figure 4.7).

The fabrication of a silicon-fusion-bonded sensor begins with the etching of a cavity in a bottom handle wafer. Silicon-fusion bonding of a second top wafer encapsulates and seals the cavity. Electrochemical etching or standard polishing thins down the top bonded wafer to form a membrane of appropriate thickness. The remaining process steps define the piezoresistive sense elements as well as the metal interconnects, and are similar to those used in the fabrication of standard, bulk micromachined pressure sensors, described earlier.

Calibration and correction of error sources are necessary for the manufacture of precision pressure sensors. A specification on accuracy of

The Gearbox: Commercial MEM Structures and Systems

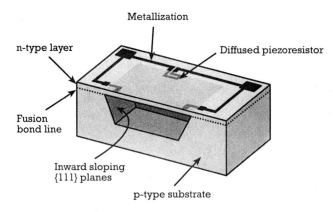

Figure 4.7 A miniature silicon-fusion-bonded absolute-pressure sensor. Courtesy of Lucas NovaSensor, Fremont, California.

better than 1% over a temperature range of −40 to 125° C is typical of many automotive, medical, and industrial applications. First-order errors include zero offset (the output at no applied pressure), uncalibrated sensitivity or span (conversion factor between input pressure and output signal), and temperature dependence of the output signal. Second-order effects include nonlinearities in the output response, as well as temperature coefficients of some first-order error terms. Compensation and correction techniques place certain restrictions on the device and process design. For example, one scheme uses laser-trimming of resistors with near-zero temperature coefficients of resistance (TCR) to correct first-order errors. But this scheme requires that the average doping concentration of the piezoresistors is above 5×10^{19} cm^{-3}, or below 3×10^{17} cm^{-3}. Compensation over intermediate ranges of dopant concentration requires sophisticated electronic circuits that continuously monitor the temperature of the Wheatstone bridge.

There has recently been a shift within the industry to provide the calibration and compensation functions with specially designed application-specific integrated circuits (ASICs). The active circuits amplify the voltage output of the piezoresistive bridge to standard CMOS voltage levels (0 to 5 V). They also correct for temperature errors and nonlinearities. Error coefficients particular to a sensor are permanently stored in on-board electrically programmable memory (e.g., EEPROM). Most sensor manufacturers have developed their own proprietary circuit designs. A few general-purpose signal-conditioning integrated circuits are commercially available; one example is the MAX1457 from Maxim Integrated Products of Sunnyvale, California.

High-temperature pressure sensors

The temperature rating of most commercially available silicon-micromachined pressure sensors is −40 to +125° C, covering the automotive and military specifications. The increased leakage current above 125° C across the *p-n* junction between the diffused piezoresistive element and the substrate significantly degrades performance. Silicon-on-insulator (SOI) technology becomes very useful at elevated temperatures because the thin silicon sense elements exist over a layer of silicon dioxide, thus eliminating all *p-n* diode junctions. Adjacent silicon sense elements are isolated from each other by shallow, moat-like trenches. The dielectric isolation below the sense elements completely eliminates the leakage current through the substrate as long as the applied voltages are below the breakdown voltage of the insulating oxide layer.

A high-temperature pressure sensor from Lucas NovaSensor utilizes SOI technology to form thin *p*-type crystalline silicon piezoresistors over a thin layer of silicon dioxide. Gold metalization and bond pads provide electrical contacts to the sense elements (Figure 4.8).

Silicon-fusion bonding plays an important role in the making of the silicon-on-insulator (SOI) substrates. A heavily doped, thin *p*-type layer is formed on the surface of one wafer, and an oxide layer is thermally grown on another wafer. Silicon fusion bonding brings the two substrates together such that the *p*-type layer is in direct contact with the oxide layer. Etching in ethylenediamine pyrocathecol (EDP) thins down the stack and stops on the heavily doped *p*-type silicon. A front-side lithography step followed by a silicon etch patterns the piezoresistive sense elements. Gold metalization is sputtered or evaporated, and then lithographically patterned to form electrical interconnects and bond pads. The final step forms a thin membrane by etching a cavity from the back side using potassium hydroxide or a similar etch solution. Double-sided lithography is critical to align the cavity outline on the back side with the piezoresistors on the front side. The front side does not need to be protected during the etch of the cavity if EDP is used instead of potassium hydroxide; EDP is highly selective to heavily doped *p*-type silicon, silicon

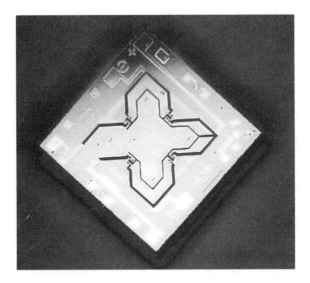

Figure 4.8 Photograph of a silicon-on-insulator-based pressure sensor rated for extended temperature operation up to 300° C. Courtesy of Lucas NovaSensor, Fremont, California.

dioxide, and gold, but it is toxic and must be handled with extreme caution (Figure 4.9).

Mass flow sensors

The flow of gas over the surface of a heated element produces convective heat loss at a rate proportional to mass flow. Flow sensors operating on this principle belong to a general category of devices known as "hot-wire anemometers." They measure the temperature of the hot element, and infer the flow rate. A number of demonstrations exist in the open literature. They all share a basic structure consisting of a thin-film heating element and a temperature-measuring device on a thin ($< 1\ \mu m$) insulating dielectric membrane suspended over an etched cavity, at least $50\ \mu m$ in depth. This architecture provides excellent thermal isolation between the heater and the supporting mechanical frame, which ensures that heat loss is nearly all due to mass flow over the heating element. A thermal isolation exceeding 15° C per milliwatt of heater power is often typical. Moreover, the small heat capacity due to the tiny, heated volume provides a fast thermal time constant, and consequently a rapid response

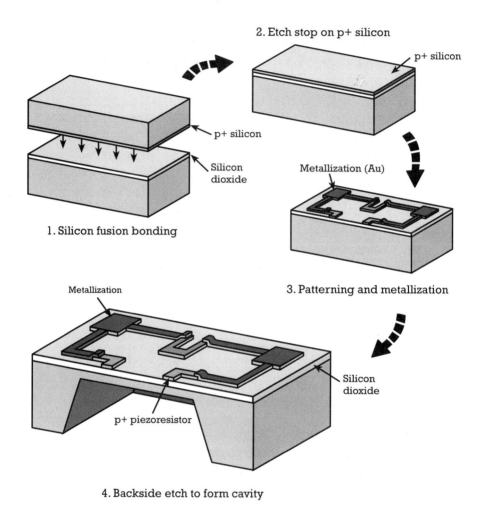

Figure 4.9 Fabrication process of a SOI high-temperature pressure sensor. Courtesy of Lucas NovaSensor, Fremont, California.

time. One approach to inferring the temperature of the heating element is to measure its resistance, and calculate the temperature using the temperature coefficient of resistance (TCR). Alternatively, direct temperature measurement using a *p-n* diode or a thermocouple is equally adequate.

Honeywell, Inc., of Minneapolis, Minnesota manufactures the AWM series of bidirectional mass-airflow sensors using two adjacent thin membranes, presumably made of silicon nitride, each containing a heating element and a temperature-sensitive resistor [8]. The two membranes are small in size, each measuring less than 500 × 500 μm^2. Gas flow across the membranes cools the upstream heater, and heats the downstream element. The two heaters are part of a first Wheatstone bridge, and the temperature-sensing resistors form two legs in a second Wheatstone bridge whose differential output is directly proportional to the rate of flow. The direction of flow is reflected in the polarity of the differential bridge output—a characteristic of the dual sense element configuration. In essence, this polarity determines which of the two heaters is upstream and which is downstream. Laser-trimmed thick- or thin-film resistors provide calibration as well as nulling of any offsets due to resistance mismatch in the Wheatstone bridges. The Honeywell AWM series of devices is capable of measuring flow rates in the range of 0 to 1,000 sccm. The upper limit is due to pronounced nonlinear effects in the heat-transfer mechanism. The full-scale output is approximately 75 mV, and the response time is less than 3 ms. The device consumes less than 30 mW (Figure 4.10).

While the processing details of the Honeywell series of airflow sensors are not publicly disclosed, one can readily design a process for fabricating a demonstration-type device. An example process would begin with the deposition of a thin layer of silicon nitride, approximately 0.5 μm in thickness, over a {100} silicon wafer. Silicon nitride is usually an excellent choice for making thin membranes because it can be deposited under tensile stress, and it retains its structural integrity in most anisotropic etch solutions. The thin-film heaters and sense elements are deposited next by sputtering a thin metal layer; for example, platinum or nickel; or by the chemical vapor deposition of a heavily doped layer of polysilicon. The thin metal film or polysilicon is then patterned using standard lithography, followed by an appropriate etch step. An insulating passivation layer, preferably made of silicon nitride, encapsulates and protects the heating and sense elements. Both silicon nitride layers must then be lithographically patterned in the shape of the two suspended membranes, and consequently etched to expose the silicon regions outside of the membrane outline. The final step involves the etching of the silicon in potassium hydroxide, or a similar anisotropic etch solution, to form the deep cavity. The etch first proceeds in the open silicon regions, and then it progresses underneath the silicon nitride thin-film, removing all the silicon and resulting in the suspended silicon nitride membranes. The reason the etch proceeds underneath the silicon nitride layer is because its

108   An Introduction to Microelectromechanical Systems Engineering

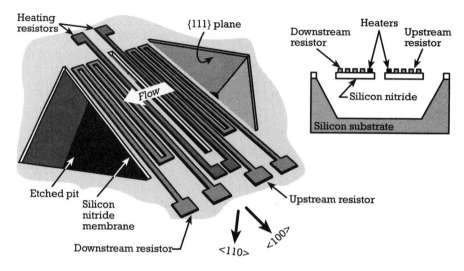

Figure 4.10 Illustration of a micromachined-mass-flow sensor. Gas flow cools the upstream heater and heats the downstream heater. Temperature-sensitive resistors are used to measure the temperature of each heater and consequently infer the flow rate. The etched pit underneath the heater provides exceptional thermal isolation to the silicon support frame. Adapted from technical sheets on the AWM series of mass-airflow sensors (Honeywell, Inc., Minneapolis, Minnesota), and from Johnson and Higashi [8].

orientation is in the <100> direction. The etch stops on the {111} crystallographic planes along the periphery of the open silicon areas.

Acceleration sensors

The first demonstration of a micromachined accelerometer took place in 1979 at Stanford University, but it took nearly 15 years before such devices became accepted mainstream products for large-volume applications. The overall unit-volume market for silicon microaccelerometers has been steadily increasing, reaching an estimated 27 million units in 1998, and driven primarily by the need for crash-sensing in airbag deployment systems. The increase in unit volume has been accompanied by a steady decrease in pricing for automotive applications, from an estimated $10 per unit in the early 1990s to about $3 per unit in 1998. Clearly, small-volume pricing for custom designs remains well above quoted figures for the high-volume automotive markets (Table 4.3).

Table 4.3
Some Applications for Micromachined Accelerometers

Measurement	Application
Acceleration	Front and side airbag crash sensing
	Electrically controlled car suspension
	Safety belt pretensioning
	Vehicle and traction control systems
	Inertial measurement, object positioning, and navigation
	Human activity for pacemaker control
Vibration	Engine management
	Condition-based maintenance of engines and machinery
	Security devices
	Shock and impact monitoring
	Monitoring of seismic activity
Angles of inclination	Inclinometers and tilt sensing
	Vehicle stability and roll
	Computer peripherals (e.g., joystick, head mounted displays ...)
	Handwriting recognition (e.g., SmartQuill from British Telecom)
	Bridges, ramps, and construction

All accelerometers share a basic structure consisting of an inertial mass suspended from a spring (Figure 4.11). However, they differ in the sensing of the relative position of the inertial mass as it displaces under the effect of an externally applied acceleration. A common sensing method is capacitive where the mass forms one side of a two-plate capacitor. This approach requires the use of special electronic circuits to detect minute changes in capacitance ($< 10^{-15}$ F) and to translate them into an amplified output voltage. Another common method uses piezoresistors to sense the internal stress induced in the spring. In yet a different method, the spring is piezoelectric or contains a piezoelectric thin-film, providing a voltage in direct proportion to the displacement. In some rare instances, such as in operation at elevated temperatures, position-sensing with an optical fiber becomes necessary. The focus of this section is on capacitive and piezoresistive accelerometers.

The primary specifications of an accelerometer are range, often given in G, the earth's gravitational acceleration (1 G = 9.81 m/s^2); sensitivity (V/G); resolution (G); bandwidth (Hz); cross-axis sensitivity; and

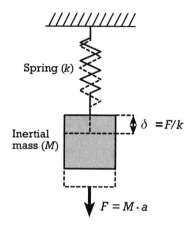

Figure 4.11 The basic structure of an accelerometer, consisting of an inertial mass suspended from a spring. The resonant frequency and the noise-equivalent acceleration (due to Brownian noise) are given.

immunity to shock. The range and bandwidth required vary significantly, depending on the application. Accelerometers for airbag crash-sensing are rated for a full range of ±50 G and a bandwidth of about one kilohertz. In contrast, devices for measuring engine knock or vibration have a range of about one G, but must resolve small accelerations (< 100 μG) over a large bandwidth (> 10 kHz). Modern cardiac pacemakers incorporate multiaxis accelerometers to monitor the level of human activity, and correspondingly adjust the stimulation frequency. The ratings on such sensors are ±2 G and a bandwidth of less than 50 Hz, but they require extremely low power consumption for battery longevity. Accelerometers for military applications can often exceed a rating of 1,000 G.

Cross-axis sensitivity assesses the immunity of the sensor to accelerations along directions perpendicular to the main sensing axis. Cross-axis rejection ratios in excess of 40 dB are always desirable. Shock immunity is an important but somewhat subjective specification for the protection of the devices during handling or operation. While one would expect the specification to be quantified in units of acceleration, it is instead defined in terms of a peculiar, but more practical test involving dropping the device from a height of one meter over concrete—the shock impact can easily reach a dynamic peak of 10,000 G! In addition to achieving a large impact, the drop test excites the various modes of resonance that are liable to cause catastrophic failure.

The Gearbox: Commercial MEM Structures and Systems

While many companies offer micromachined acceleration sensor products, a representative set of only four accelerometers follows next, each unique in its design and fabrication.

Piezoresistive bulk-micromachined accelerometer

Until recently, piezoresistive-type acceleration sensors were widely available. Many companies, including Lucas NovaSensor, Fremont, California; and EG&G IC Sensors, Milpitas, California, offered products using an anisotropically etched silicon-inertial-mass and diffused-piezoresistive-sense elements. But these products were retired because they could not meet the aggressive pricing requirements of the automotive industry. The product introduction in 1996 by Endevco Corporation of San Juan Capistrano, California indicates that piezoresistive accelerometers remain in this highly competitive market (Figure 4.12).

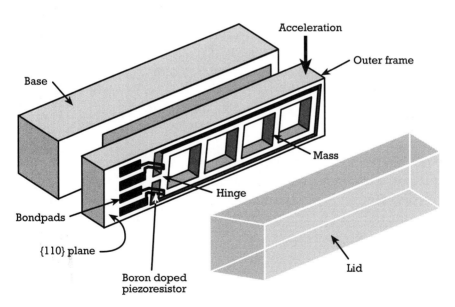

Figure 4.12 Illustration of a piezoresistive accelerometer from Endevco Corporation fabricated using anisotropic etching in {110} wafers. The middle core contains the inertial mass suspended from a hinge. Two piezoresistive sense elements measure the deflection of the mass. The axis of sensitivity is in the plane of the middle core. The outer frame acts as a stop mechanism to prevent excessive accelerations from damaging the part. Adapted from J. T. Suminto [9].

The Endevco sensor consists of three substrates: a lower base, a middle core containing a hinge-like spring, the inertial mass, and the sense elements; and finally a top protective lid [9]. The inertial mass sits inside a frame suspended by the spring. Two thin, boron-doped piezoresistive elements in a Wheatstone bridge configuration span the narrow 3.5-μm-gap between the outer frame of the middle core and the inertial mass. The piezoresistors are only 0.6-μm-thick and 4.2-μm-long, and are thus very sensitive to minute displacements of the inertial mass. The output in response to an acceleration equal to one G in magnitude is 25 mV for a Wheatstone bridge excitation of 10 V. The thick and narrow hinge structure allows displacement within the plane of the device, but it is very stiff in directions normal to the wafer, resulting in high immunity to off-axis accelerations. Moreover, the outer frame acts as a stop mechanism protecting the device in the event of excessive acceleration shocks. It takes 6,000 G for the inertial mass to touch the frame, and the device can survive shocks in excess of 10,000 G. Open apertures reduce the weight of the inertial mass, and combine with the stiff hinge to provide a rather high resonant frequency of 28 kHz.

The fabrication process is somewhat unique with its utilization of {110} wafers for the middle core. In this case, some {111} crystallographic planes are orthogonal to the {110} surface of the wafer, which allows the formation of vertical trenches using anisotropic wet etchants. The fabrication begins with boron implantation and diffusion at 1100° C to form highly doped p-type piezoresistors. In order to obtain maximum sensitivity, the piezoresistors are aligned along a <111> direction. A silicon oxide or silicon nitride layer masks the silicon in the form of the inertial mass and hinge during the subsequent anisotropic etch in ethylenediamine pyrocathechol (EDP). The inertial mass is bounded by vertical {111} planes, giving it the shape of a parallelogram whose inside angle is 70.5° (see Chapter 3). Subsequent fabrication steps provide for the deposition and patterning of aluminum electrical contacts and bond pads. Shallow recesses are incorporated into the base and lid substrates before the three-wafer stack is bonded together, using solder glass as adhesive.

Capacitive bulk-micromachined accelerometer

Many companies offer capacitive bulk-micromachined accelerometers. The next example describes the SCA series from VTI Hamlin, Vantaa,

The Gearbox: Commercial MEM Structures and Systems

Finland; a subsidiary of Breed Technologies, Inc., Lakeland, Florida. The sensor consists of a stack of three bonded silicon wafers, with the hinge spring and inertial mass incorporated into the middle wafer. The inertial mass forms a movable inner electrode of a variable differential capacitor circuit. The two outer wafers are identical and are simply the fixed electrodes of the two capacitors (Figure 4.13).

Holes through the inertial mass reduce the damping effect from air trapped in the enclosed cavity, increasing the operating bandwidth of the sensor. Unlike other designs, the contacts to the electrodes are on the side of the die, and thus must be defined after the wafer is diced into individual sensor parts. The SCA series of sensors is available in a measuring range from ±1.5 G to ±50 G. Electronic circuits sense changes in capacitance, then convert them into an output voltage between ±2.5 V, with the sign indicating the direction of the acceleration vector. The rated bandwidth is

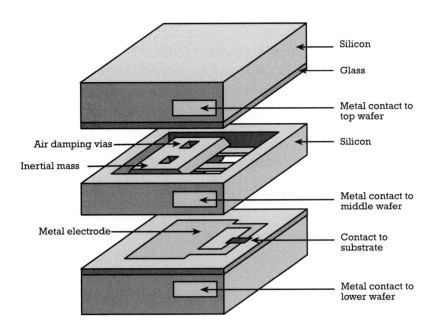

Figure 4.13 Illustration of a capacitive bulk-micromachined accelerometer. The inertial mass in the middle wafer forms the movable electrode of a variable differential capacitive circuit. Adapted from the accelerometer product catalog (VTI Hamlin, Vantaa, Finland).

400 Hz, the cross-axis sensitivity is less than 5% of output, and the shock immunity is 20,000 G.

The particulars of the VTI Hamlin process are not publicly available. However, Sasayama et al. [10] describe a process for building a similar part. The three wafers are fabricated separately, then joined at the end by a bonding process, such as anodic bonding, silicon-fusion bonding, or even a glass thermocompression bond. The upper and lower wafers are identical, and contain a metal electrode patterned with standard lithography over a thin layer of silicon dioxide. The inertial mass and hinge are delineated in the middle wafer using four sequential steps of anisotropic etching in potassium hydroxide, or a similar etchant. First, shallow recess cavities are etched on both sides of the wafer. Three distinct masking layers are then each deposited and patterned separately. Silicon dioxide and silicon nitride are good material choices. Each of these masking layers is sequentially removed after an etch step in an anisotropic wet-etching solution. In essence, the pattern information is encoded in each of the three masking layers. Timed etching simply translates the encoded information into a variable topography in the silicon substrate. The end result is a thin support hinge member with a much thicker inertial mass. The recesses on either side of the mass form the thin gaps for the two-plate sense capacitors (Figure 4.14).

Capacitive surface-micromachined accelerometer

Surface micromachining emerged in the late 1980s as a perceived low-cost alternative for accelerometers, aimed primarily at automotive applications. Both Robert Bosch GmbH, Stuttgart, Germany; and Analog Devices, Inc., Norwood, Massachusetts; offer surface-micromachined accelerometers, but it is the latter company that benefited from wide publicity of their ADXL product family [11]. The Bosch sensor [12] is incorporated in the Mercedes Benz family of luxury automobiles. The ADXL part is used in Ford, General Motors, and other vehicles, as well as inside joysticks for computer games. The surface-micromachining fabrication sequence, illustrated in Chapter 3, is fundamentally similar to both sensors, though the Bosch device uses a thicker ($10\,\mu$m) polysilicon structural element.

Unlike most bulk-micromachined parts, surface-micromachined accelerometers incorporate a suspended comb-like structure whose primary axis of sensitivity lies in the plane of the die. This is often referred to

The Gearbox: Commercial MEM Structures and Systems

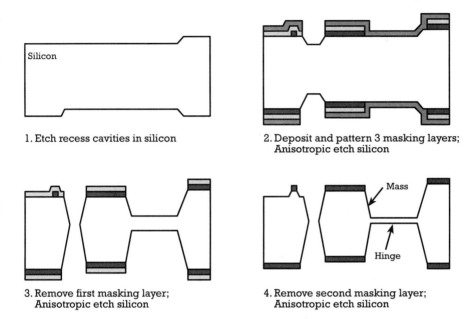

Figure 4.14 Process steps to fabricate the middle wafer containing the hinge and the inertial mass of a capacitive bulk-micromachined accelerometer similar to the device from VTI Hamlin. Adapted from Sasayama et al. [10].

as an *x-axis* (or *y-axis*)-type device, as opposed to *z-axis* sensors, where the sense axis is orthogonal to the plane of the die. However, due to the relative thinness of their structural elements, surface micromachined accelerometers suffer from sensitivity to accelerations out of the plane of the die (*z*-axis). Shocks along this direction can cause catastrophic failures.

The ADXL device [11] consists of three sets of 2-μm-thick polysilicon finger-like electrodes (Figure 4.15). Two sets are anchored to the substrate and are stationary. They respectively form the upper and lower electrode plates of a differential capacitance system. The third set has the appearance of a two-sided comb whose fingers are interlaced with the fingers of the first two sets. It is suspended approximately 1 μm over the surface by means of two long and folded polysilicon beams acting as suspension springs. It also forms the common middle and displaceable electrode for the two capacitors. The inertial mass consists of the comb fingers and the central backbone element to which these suspended

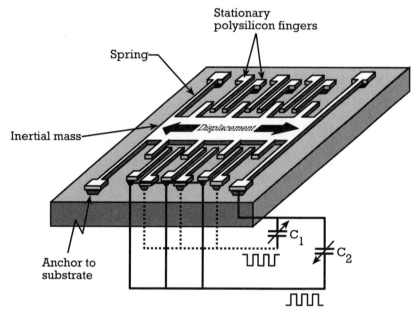

Figure 4.15 Illustration of the basic structure of the ADXL family of surface-micromachined accelerometers. A comb-like structure suspended from springs forms the inertial mass. Displacements of the mass are measured capacitively with respect to two sets of stationary finger-like electrodes. Adapted from the ADXL data sheets and application notes (Analog Devices, Inc., Norwood, Massachusetts).

fingers are attached. Under no externally applied acceleration, the two capacitances are identical. The output signal, proportional to the difference in capacitance, is null. An applied acceleration displaces the suspended structure, resulting in an imbalance in the capacitive half-bridge. The differential structure is such that one capacitance increases and the other decreases. The overall capacitance is small, typically on the order of 100 fF (1 fF = 10^{-15} F). For the ADXL05 (rated at ±5 G), the change in capacitance in response to one G is minute, about 100 aF (1 aF = 10^{-18} F). This is equivalent to only 625 electrons at an applied bias of one volt, and thus must be measured using on-chip integrated electronics to greatly reduce the impact of parasitic sources. The basic read-out circuitry consists of a small amplitude, two-phase oscillator driving both ends of the capacitive half-bridge in opposite phases at a frequency of 1 MHz. A capacitance imbalance gives rise to a voltage in the middle node. The signal is then demodulated and amplified. The one-MHz excitation frequency is sufficiently higher than the mechanical resonant frequency so that it produces no actuation force on the plates of the capacitors, provided its DC (average) value is null. The maximum acceleration rating for the ADXL family varies from ±2 G (ADXL 202) up to ±100 G (ADXL 190). The dynamic range is limited to about 60 dB over the operational bandwidth (typically, 1 to 6 kHz). The small change in capacitance and the relatively small mass combine to give a noise floor that is relatively large, when compared to similarly rated bulk-micromachined or piezoelectric accelerometers. For the ADXL05, the mass is only 0.3 μg, and the corresponding noise floor, dominated by Brownian mechanical noise, is 500 $\mu G/\sqrt{Hz}$. By contrast, the mass for a bulk-micromachined sensor can easily exceed 100 μg.

Applying a large-amplitude voltage at low frequency—below the natural frequency of the sensor—between the two plates of a capacitor gives rise to an electrostatic force that tends to pull the two plates together. This effect enables applying feedback to the inertial mass: Every time the acceleration pulls the set of suspended fingers away from one of the anchored sets, a voltage significantly larger in amplitude than the sense voltage, but lower in frequency, is applied to the same set of plates, pulling them together and effectively counterbalancing the action of the external acceleration. This feedback voltage is appropriately proportioned to the measured capacitive imbalance in order to maintain the suspended fingers in their initial position, practically in a pseudostationary state. This electrostatic actuation, also called force balancing, is a form of closed-loop feedback. It minimizes displacement, and greatly improves output linearity since the center element never quite moves by more than a few nanometers. The sense and actuation plates may be the same provided the two frequency signals (sense and actuation) do not interfere with one another.

A significant advantage of surface micromachining is the ease of integrating two single-axis accelerometers on the same die to form a dual-axis accelerometer, so-called *two-axes*. In a very simple configuration, the two accelerometers are orthogonal to each other. However, the ADXL200 series of dual-axis sensors employs a more sophisticated suspension spring mechanism where a single inertial mass is shared by both accelerometers.

Capacitive deep-etched micromachined accelerometer

The deep-reactive-ion-etched (DRIE) accelerometer from Lucas NovaSensor, Fremont, California, shares its basic comb structure design with the ADXL and Bosch accelerometers. It consists of a set of fingers attached to a central backbone plate, itself suspended by two folded springs. Two sets of stationary fingers attached directly to the substrate complete the capacitive half-bridge (Figure 4.16). The design, however, adds a few improvements. By taking advantage of the third dimension and using structures 50- to 100-μm-deep, the sensor gains a larger inertial mass, up to 100 μg, as well as a larger capacitance, up to 5 pF. The relatively large mass reduces mechanical Brownian noise, and increases resolution. The high-aspect ratio of the spring practically eliminates the sensitivity to z-axis accelerations (out of the plane of the die). Fabrication follows the silicon-fusion-bondingdeep—reactive etching (SFB-DRIE) process introduced in Chapter 3.

The sensor, described by van Drieënhuizen, et al. [13], uses a 60-μm-thick comb structure for a total capacitance of 3 pF, an inertial mass of 43 μg, a resonant frequency of 3.1 kHz, and an open-loop mechanical sensitivity of 1.6 fF/G. The corresponding mechanical noise is

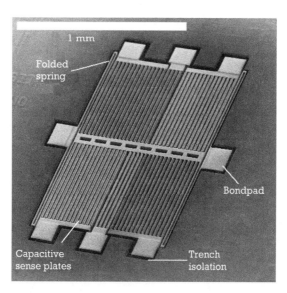

Figure 4.16 Photograph of a deep-reactive-ion-etched (DRIE) accelerometer using 60-μm-thick comb structures. Courtesy of Lucas NovaSensor, Fremont, California.

about 10 $\mu G/\sqrt{Hz}$, significantly less than for a surface-micromachined sensor. The read-out circuitry first converts changes in capacitance into frequency. This is accomplished by inserting the two variable capacitors into separate oscillating circuits whose output frequencies are directly proportional to the capacitance. A phase detector compares the two output frequencies and converts the difference into a voltage. The circuit then amplifies the signal before feeding it back to a set of actuation electrodes for force balancing. These electrodes may be distinct from the sense electrodes. Filters set the closed-loop bandwidth to 1 kHz. The overall sensitivity is 700 mV/G for a ±5 G device. Early prototypes had a dynamic range of 44 dB, limited by electronic $1/f$ noise in the CMOS circuitry. Recent prototypes, with newer implementations of the electronic read-out circuits, demonstrated a dynamic range approaching 70 dB over the 1-kHz bandwidth. The SFB-DRIE process is fully compatible with the integration of CMOS circuits next to the mechanical sensing element. The large available capacitance makes the decision to integrate based purely on economics, rather than on performance.

Angular rate sensors and gyroscopes

Long before the advent of Loran and the satellite-based GPS system, the gyroscope was a critical navigational instrument used for maintaining a fixed orientation with great accuracy, regardless of earth rotation. Invented in the 19th century, it consisted of a flywheel mounted in gimbal rings (Figure 4.17). The large angular momentum of the flywheel counteracts externally applied torques, and keeps the orientation of the spin-axis unaltered. The demonstration of the ring-laser gyroscope in 1963 displaced the mechanical gyroscope in many high-precision applications, including aviation. Inertial navigation systems based on ring-laser gyroscopes are on board virtually all commercial aircraft. Gyroscopes capable of precise measurement of rotation are very expensive instruments, costing many thousands of dollars. An article, published in 1984 by the IEEE, reviews many of the basic technologies for gyroscopes [14].

The gyroscope derives its precision from the large angular momentum that is proportional to the heavy mass of the flywheel, and its substantial size and high rate of spin. This, in itself, precludes the use of miniature devices for useful gyroscopic action; the angular momentum of

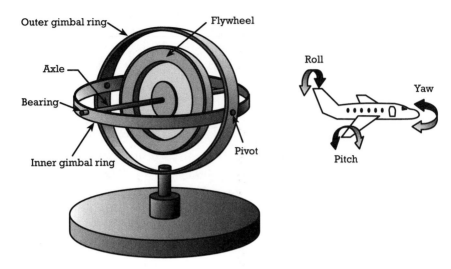

Figure 4.17 Illustration of a conventional mechanical gyroscope and the three rotational degrees of freedom it can measure.

a miniature flywheel is miniscule. Instead, micromachined sensors that detect angular rotation utilize the Coriolis effect. Fundamentally, such devices are strictly angular-rate or yaw-rate sensors, measuring angular velocity. However, they are colloquially but incorrectly referred to as gyroscopes.

The Coriolis effect, named after the French physicist Gaspard Coriolis, manifests itself in numerous weather phenomena including hurricanes and tornadoes, and is a direct consequence of a body's motion in a rotating frame of reference. To understand it, let us imagine an automobile driving from Seattle, Washington (lat. 48° N), to Los Angeles, California (lat. 34° N). At the beginning of its journey, the car in Seattle is actually moving eastward with the rotation of Earth (the rotating frame of reference), at about 1120 km/h[1]. At the end of its journey in Los Angeles, its eastward velocity is 1385 km/h. As the car moves south across latitudes, its eastward velocity must increase from 1120 to 1385 km/h, otherwise it will continuously slip and never reach its destination. The road—effectively the rotating surface—imparts an eastward acceleration to maintain the vehicle on its course. This is the Coriolis acceleration. In general, the Coriolis acceleration is the acceleration that must be applied in

---

1. The velocity at the equator is 1670 km/h. The velocity at latitude 48° N is 1670 km/h multiplied by cos 48°.

order to maintain the heading of a body moving on a rotating surface [15 (Figure 4.18)].

All micromachined angular-rate sensors have a vibrating element at their core—this is the moving body. In a fixed frame of reference, a point on this element oscillates with a velocity vector v. If the frame of reference begins to rotate at a rate Ω, this point is then subject to a Coriolis force and a corresponding acceleration equal to $2\Omega \times v$ [16]. The vector cross-operation implies that the Coriolis acceleration and the resulting displacement at that point are perpendicular to the oscillation. This, in effect, sets up an energy-transfer process from a primary mode of oscillation into a secondary mode that can be measured. It is this excitation of a secondary resonance mode that forms the basis of detection using the Coriolis effect. In beam structures, these two frequencies are distinct with orthogonal displacements. But for highly symmetrical elements such as rings, cylinders, or disks, the resonant frequency is degenerate, meaning there are two distinct modes-of-resonance sharing the same oscillation frequency. This degeneracy causes the temporal excitation signal (primary mode) to be in phase-quadrature with the sense signal (secondary

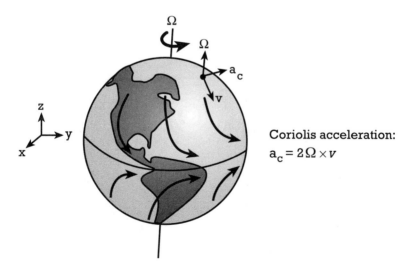

Figure 4.18 Illustration of the Coriolis acceleration on an object moving with a velocity vector v on the surface of Earth from either pole towards the equator. The Coriolis acceleration deflects the object in a counterclockwise manner in the northern hemisphere and a clockwise direction in the southern hemisphere. The vector Ω represents the rotation of the planet.

mode), thus minimizing coupling between these two modes, and improving on sensitivity and accuracy [17]. In addition, the degeneracy tends to minimize the device's sensitivity to thermal errors, aging, and long-term frequency drifts.

A simple and common implementation is the tuning fork structure (Figure 4.19). The two tines of the fork normally vibrate in opposite directions in the plane of the fork (flexural mode). The Coriolis acceleration subjects the tips to a displacement perpendicular to the primary mode of oscillation, forcing each tip to describe an elliptical path. Rotation, hence, excites a secondary torsional vibration mode about the stem, with energy transferred from the primary flexural vibration of the tines. Quartz tuning forks such as those from Systron Donner, Concord, California, use the piezoelectric properties of the material to excite and sense both vibration modes. The tuning fork structure is also at the core of a micromachined silicon sensor from Daimler Benz AG, that will be described later. Other implementations of angular-rate sensors include simple resonant beams, vibrating ring shells, and tethered accelerometers; but all of them exploit the principle of transfering energy from a primary to a secondary mode of resonance. Of all the vibrating angular-rate structures, the ring shell or cylinder is the most promising for inertial- and navigational-grade performance because of the frequency degeneracy of its two resonant modes.

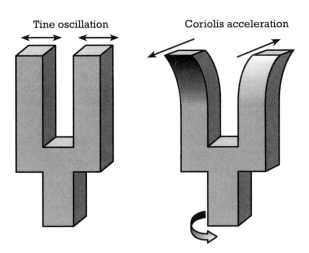

Figure 4.19 Illustration of the tuning fork structure for angular-rate sensing. The Coriolis effect transfers energy from a primary, flexural mode to a secondary, torsional mode.

The Gearbox: Commercial MEM Structures and Systems

The main specifications of an angular-rate sensor are full-scale range, expressed in °/s or °/hr; scale factor or sensitivity [V/(°/s)]; noise also known as angle random walk [°/(s · $\sqrt{Hz}$)]; bandwidth (Hz); resolution (°/s); and dynamic range (dB); the latter two being functions of noise and bandwidth. Short- and long-term drift of the output, known as bias drift, is another important specification expressed in °/s or °/hr. As is the case for most sensors, angular-rate sensors must withstand shocks of at least 1,000 G.

Micromachined angular-rate sensors have largely been unable to deliver a performance better than "rate grade." These are devices with a dynamic range of only 40 dB, a noise figure larger than 0.1 °/(s · $\sqrt{Hz}$), and a bias drift worse than 10 °/hr. By comparison, "inertial grade" sensors and true gyroscopes deliver a dynamic range of over 100 dB, a noise less than 0.001 °/(hr · $\sqrt{Hz}$), and a bias drift better than 0.01 °/hr [18]. But the advantage of micromachined angular-rate sensors lies in their small size and low cost, currently less than $20. They are slowly gaining acceptance in automotive applications, in particular, for vehicle stability systems. The sensor detects any undesired yaw of a vehicle due to poor road conditions, and feeds the information to a control system, which may activate the anti-lock braking system (ABS) or the traction control system (TCS) to correct the situation. The Mercedes Benz ML series of sport utility vehicles incorporates a silicon angular-rate sensor from Robert Bosch GmbH for vehicle stability.

The selection of commercially available micromachined yaw-rate sensors remains limited, but many manufacturers have publicly acknowledged the existence of development programs. The sensors from Delco Electronic Corporation, Robert Bosch GmbH, Daimler Benz AG, and British Aerospace Systems and Equipment, illustrate four vibratory type angular-rate sensors distinct in their structure as well as excitation and sense methods.

Micromachined angular-rate sensor from Delco Electronics

The sensor from Delco Electronics Corporation, Kokomo, Indiana [19], includes at its core a vibrating ring shell, based on the principle of the ringing wine glass, discovered in 1890 by G. H. Bryan (Figure 4.20). He observed that the standing wave pattern of the wine glass did not remain stationary in inertial space, but participated in the motion as the glass rotated about its stem.

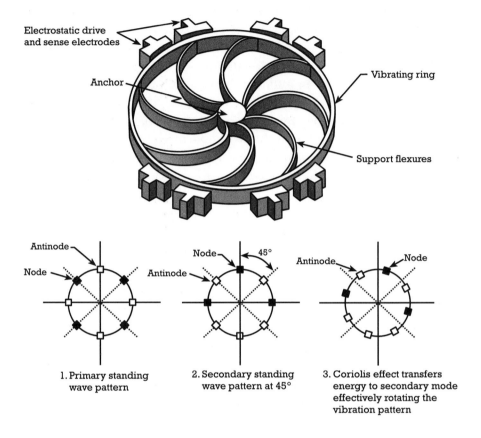

Figure 4.20 Illustration of the Delco angular-rate sensor and the corresponding standing wave pattern. The basic structure consists of a ring shell suspended from an anchor by support flexures. A total of 32 electrodes (only a few are shown) distributed around the entire perimeter of the ring excite a primary mode of resonance using electrostatic actuation. A second set of distributed electrodes capacitively sense the vibration modes. The angular shift of the standing wave pattern is a measure of the angular velocity. Adapted from Chang et al. [19].

The complete theory of vibrating ring angular-rate sensors is well developed [20]. The ring shell, anchored at its center to the substrate, deforms as it vibrates through a full cycle from a circle to an ellipse, back to a circle, then to an ellipse rotated at right angles to the first ellipse, then back to the original circle. The points on the shell that remain stationary are called nodes, whereas the points that undergo maximal deflection are called antinodes. The nodes and antinodes form a vibration pattern—or standing wave pattern—around the ring that is characteristic of the resonance mode. Because of symmetry, a ring shell possesses two frequency-degenerate resonant modes, with their vibration patterns offset by 45° with respect to each other. Hence, the nodes of the first mode coincide with the antinodes of the second mode. The external control electronics excite only one of the two modes—the primary mode. But under rotation, the Coriolis effect excites the second resonance mode, and energy transfer occurs between the two modes. Consequently, the deflection amplitude builds up at the antinodes of the second mode—also, the nodes of the first mode. The overall vibration becomes a linear combination of the two modes, with a new set of nodes and antinodes forming a vibration pattern rotated with respect to the pattern of the primary mode. It is this lag that Bryan heard in his spinning wine glass. In an open-loop configuration, the deflection amplitude at the nodes and antinodes is a measure of the angular rate of rotation. Alternatively, the angular shift of the vibration pattern is another measure. In a closed-loop configuration, electrostatic actuation by a feedback voltage applied to the excitation electrodes nulls the secondary mode and maintains a stationary vibration pattern. The angular rate becomes directly proportional to this feedback voltage.

A total of 32 electrodes positioned around the suspended ring shell provide the electrostatic excitation drive and sense functions. Of this set, eight electrodes strategically positioned at 45° intervals—at the nodes and antinodes—capacitively sense the deformation of the ring shell. Appropriate electronic circuits complete the system control functions, including feedback. A phased-locked loop (PLL) drives the ring into resonance through the electrostatic drive electrodes, and maintains a lock on the frequency. Feedback is useful to electronically compensate for the mechanical poles and increase the closed-loop bandwidth of the sensor. In addition, a high mechanical quality factor increases the closed-loop system gain and sensitivity.

The fabrication process is similar to the electroplating and molding process described in Chapter 3, except that the substrate includes preprocessed CMOS control circuitry. The mold is made of photoresist, and the electroplated nickel ring shell is 15- to 50-μm-thick. Finally, packaging is completed in a vacuum in order to minimize air damping of the resonant ring and provide a large quality factor. Researchers at the University of Michigan demonstrated a polysilicon version of the sensor, with improved overall performance.

The demonstrated specifications of the Delco sensor, over the temperature range of −40 to +125° C, include a resolution of 0.5 °/s over a bandwidth of 25 Hz, limited by noise in the electronic circuitry. The nonlinearity in a rate range of ±100 °/s is less than 0.2 °/s. The sensor survives the standard automotive shock test: a drop from a height of one meter. The specifications are adequate for most automotive and consumer applications.

Angular-rate sensor from British Aerospace Systems and Equipment

The VSG family of yaw-rate sensors from British Aerospace Systems and Equipment, Plymouth, Devon, England, in collaboration with Sumitomo Precision Products Company, Japan, is aimed at commercial and automotive applications. It also uses a vibratory ring shell similar to the sensor from Delco, but differs in the excitation and sense methods. Electric current loops in a magnetic field, instead of electrostatic electrodes, excite the primary mode of resonance. These same loops provide the sense signal to detect the angular position of the vibration pattern (Figure 4.21).

The ring, 6 mm in diameter, is suspended by eight flexural beams anchored to a 10-mm square frame. Eight equivalent current loops span every two adjacent support beams. A current loop starts at a bond pad on the frame, traces a support beam to the ring, continues on the ring for one-eighth of the circumference, then onto the next adjacent support beam before ending on a second bond pad. Under this scheme, each support beam carries two conductors. A Samarium-Cobalt permanent magnet mounted inside the package provides a magnetic field. Electromagnetic interaction between current in a loop and the magnetic field induces a Lorentz force. Its radial component is responsible for the oscillation of the ring in the plane of the die at approximately 14.5 kHz—the mechanical resonant frequency of the ring. The sensing mechanism measures the voltage induced around one or more loops in accordance with Faraday's law: As the ring oscillates, the current loop sweeps an area through the magnetic flux, generating an electromotive force (emf). Two loops, diametrically opposite, perform a differential voltage measurement. One can simplistically view an actuating and a sensing loop as the

The Gearbox: Commercial MEM Structures and Systems

primary and secondary windings of a transformer; the electromagnetic coupling between them depends on the ring vibration pattern, and thus on the angular rate of rotation.

Closed-loop feedback improves the overall performance by increasing the bandwidth and reducing the system's sensitivity to physical errors. Two separate feedback loops with automatic gain control circuits maintain a constant oscillation amplitude for the primary mode of resonance and a zero amplitude for the secondary resonance mode. The feedback voltage required to null the secondary mode is a direct measure of the rate of rotation.

The fabrication of the sensor is relatively simple (Figure 4.21). A silicon dioxide layer is deposited on a silicon wafer, then lithographically patterned and etched. The silicon dioxide layer serves to electrically

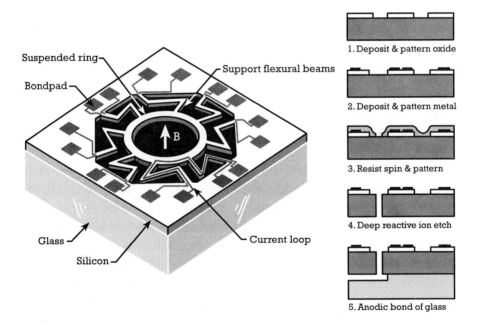

Figure 4.21 Illustration of the VSG angular-rate sensor from British Aerospace Systems and Equipment, and corresponding fabrication process. The device uses a vibratory ring shell design, similar to the Delco sensor. Eight current loops in a magnetic field, B, provide the excitation and sense functions. For simplicity, only one of the current loops is shown. Adapted from the product data sheet (British Aerospace Systems and Equipment, Plymouth, Devon, England).

isolate the current loops. A metal layer is sputter deposited, patterned, and etched to define the current loops as well as the bond pads. A layer of photoresist is spun on and patterned in the shape of the ring and support flexural beams. The photoresist then serves as a mask for a subsequent deep-reactive-ion-etching step to etch trenches through the wafer. Upon removal of the photoresist mask, the silicon wafer is anodically bonded to a glass wafer on which surface a shallow cavity was previously defined. Little is available in the open literature on the packaging, but it is clear from the need to include a permanent magnet that the packaging is custom and specific to this application.

The specification sheet of the VSG 2000 gives an output scale factor of 20 mV/(°/s) with a variation of ±2% over a temperature range from −40 to +85° C. The resolution of the sensor is 0.01 °/s, and the noise is less than 0.5 °/s over a 65-Hz bandwidth. The nonlinearity in a rate range of ±100 °/s is less than 0.2 °/s. The operating current is 100 mA at a nominal 12-V supply.

Angular-rate sensor from Daimler Benz

The sensor from Daimler Benz AG, Stuttgart, Germany [21], is a strict implementation of a tuning fork using micromachining technology. The tines of the silicon tuning fork vibrate out of the plane of the die, driven by a thin-film piezoelectric aluminum nitride actuator on top of one of the tines. The Coriolis forces on the tines produce a torquing moment about the stem of the tuning fork, giving rise to shear stresses that can be sensed with diffused piezoresistive elements. The shear stress is maximal on the center line of the stem, and corresponds with the optimal location for the piezoresistive sense elements (Figure 4.22).

The high precision of micromachining is not sufficient to ensure the balancing of the two tines and the tuning of the two resonant frequencies—recall from the discussion above that the vibration modes of a tuning fork are not degenerate. An imbalance in the tines produces undesirable coupling between the excitation and sense resonant modes, which degrades the resolution of the device. A laser ablation step precisely removes tine material, and provides calibration of the tuning fork. For this particular design, all modes of the fork are at frequencies above 10 kHz. To minimize coupling to higher orders, the primary and secondary modes are separated by at least 10 kHz from all other

The Gearbox: Commercial MEM Structures and Systems

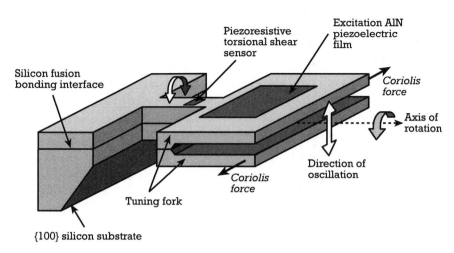

Figure 4.22 Illustration of the angular-rate sensor from Daimler Benz. The structure is a strict implementation of a tuning fork in silicon. A piezoelectric actuator excites the fork into resonance. The Coriolis force results in torsional shear stress in the stem which is measured by a piezoresistive sense element. Adapted from Voss et al. [21].

remaining modes. The choice of crystalline silicon for tine material allows for achieving a high quality factor (~ 7,000) at pressures below 10^{-5} bar.

The fabrication process is distinct from that of other yaw-rate sensors in its usage of silicon-on-insulator (SOI) substrates. The crystalline silicon over the silicon dioxide layer defines the tines. The thickness control of the tines is accomplished at the beginning of the process by the precise epitaxial growth of silicon over the SOI substrate. The thickness of the silicon layer, and consequently of the tine, varies between 20 and 200 μm, depending on the desired performance of the sensor. Lithography followed by a shallow silicon etch in tetramethyl ammonium hydroxide (TMAH) define 2-μm-deep cavities in two identical SOI substrates. Silicon-fusion bonding brings these substrates together such that the cavities are facing each other. The cavity depth determines the separation between the two tines. An etch step in TMAH removes the silicon on the front side and stops on the buried silicon dioxide layer, which is subsequently removed in hydrofluoric acid. The following steps define the piezoelectric and piezoresistive elements on the silicon surface. Diffused piezoresistors are formed using ion implantation and diffusion. Piezoelectric aluminum nitride is then deposited by sputtering aluminum in a controlled nitrogen and argon atmosphere. This layer is lithographically patterned and etched in the shape of the excitation plate over the tine. Aluminum is then sputtered and patterned to form electrical interconnects and bond pads. Finally, a TMAH etch step from the back side removes the silicon from underneath the tines. The buried silicon dioxide layer acts as an etch stop. An anisotropic plasma etch from the front side releases the tines (Figure 4.23).

The measured frequency of the primary mode (excitation mode) was 32.2 kHz, whereas the torsional secondary mode (sense mode) was 245 Hz lower. Typical of tuning forks, the frequencies exhibited a temperature dependence. For this particular technology, the temperature coefficient of frequency is –0.85 Hz/° C.

Angular-rate sensor from Robert Bosch

This sensor from Robert Bosch GmbH, Stuttgart, Germany, is unique in its implementation of a mechanical resonant structure equivalent to a tuning fork [22]. An oscillator system consists of two identical masses coupled to each other by a spring, and suspended from an outer frame by two other springs. Such a coupled system has two resonant

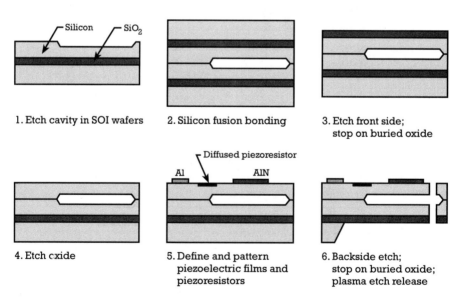

Figure 4.23 The main fabrication steps for the Daimler Benz micromachined angular-rate sensor.

The Gearbox: Commercial MEM Structures and Systems

frequencies: in-phase and out-of-phase. In the in-phase oscillation mode, the instantaneous displacements of the two masses are identical. In the out-of-phase mode, the masses are moving, at any instant, in opposite directions. A careful selection of the coupling spring provides sufficient separation between the in-phase and out-of-phase resonant frequencies. Lorentz forces generated by an electric current loop within a permanent magnetic field excite only the out-of-phase mode. The oscillation electromagnetically induces a voltage in a second current loop that provides a feedback signal proportional to the velocity of the masses. The resulting Coriolis forces on the two masses are in opposite directions but orthogonal to the direction of oscillation. Two polysilicon-surface-micromachined accelerometers with capacitive comb structures (similar in their basic operation to the ADXL family of sensors) measure the Coriolis accelerations for each of the masses. The difference of the two accelerations is a direct measure of the angular yaw rate, whereas their sum is proportional to the linear acceleration along the accelerometer's sensitive axis. Electronic circuits perform the addition and subtraction functions, and then filter out the linear acceleration signal (Figure 4.24).

For the Bosch sensor, the out-of-phase resonant frequency is 2 kHz, and the maximum oscillation amplitude at this frequency is 50 μm. The measured quality factor of the oscillator at atmospheric pressure is 1,200, sufficiently large to excite resonance with small Lorentz forces. The stimulated oscillation subjects the masses to large accelerations reaching approximately 800 G. Though they are theoretically perpendicular to the sensitive axis of the accelerometers, in practice, some coupling remains which threatens the signal integrity. However, since the two temporal signals are in-phase quadrature, adopting synchronous demodulation methods allows the circuits to filter the spurious coupled signal with a rejection ratio exceeding 78 dB. This is indeed a large rejection ratio, but insufficient to meet the requirements of inertial navigation.

The peak Coriolis acceleration for a yaw rate of 100 °/s is only 200 mG. This requires extremely sensitive accelerometers with compliant springs. The small Coriolis acceleration further emphasizes the need for perfect orthogonality between the sense and excitation axes. Closed-loop position feedback of the acceleration sense element compensates for the mechanical poles and increases the bandwidth of the accelerometers to over 10 kHz.

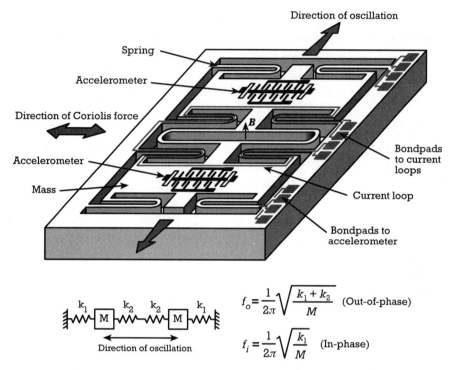

Figure 4.24 Illustration of the yaw-rate sensor from Robert Bosch GmbH. A simple mechanical model shows the two masses and coupling springs. Adapted from Lutz et al. [22].

The fabrication process simultaneously encompasses bulk and surface micromachining; the former to define the masses, and the latter to form the comb-like accelerometers (Figure 4.25). The process sequence begins by depositing a 2.5 μm layer of silicon dioxide on a silicon substrate. Epitaxy over the oxide layer grows a 12-μm-thick layer of heavily doped n-type polysilicon. This layer forms the basis for the surface-micromachined sensors, and is polycrystalline because of the lack of a seed crystal during epitaxial growth. In the next step, aluminum is deposited by sputtering and it is patterned to form electrical interconnects and bond pads. Timed etching from the back side using potassium hydroxide thins the central portion of the wafer to 50 μm. Two sequential deep-reactive-ion-etch steps define the structural elements of the accelerometers and the oscillating masses. The following step involves

The Gearbox: Commercial MEM Structures and Systems

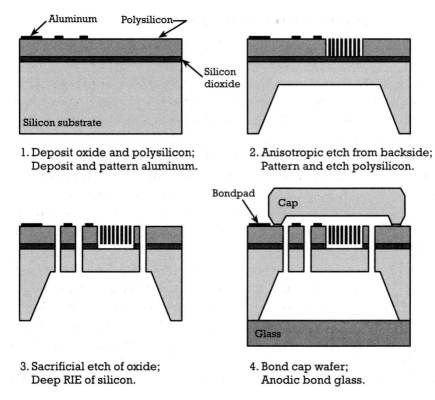

Figure 4.25 Illustration of the fabrication process for the yaw-rate sensor from Robert Bosch GmbH. Adapted from Lutz, et al. [22].

etching the sacrificial silicon dioxide layer using a gas phase process (e.g., hydrofluoric acid vapor) to release the polysilicon comb structures. Finally, a protective silicon cap wafer that contains a recess cavity is bonded on the front side using a low-temperature seal glass process. A glass wafer anodically bonded to the back side seals the device. The final assembly brings together the silicon sensor and the electronic circuits inside a metal can whose cover holds a permanent magnet.

The sensitivity of the device is 18 mV/(°/s) in the range of ±100 °/s and over −40 to +85° C. The temperature dependence of the uncompensated sensor causes an offset amplitude of 0.5 °/s over the specified temperature range, but signal-conditioning circuits reduce this dependence by implementing appropriate, electronic temperature-compensation schemes.

Radiation sensors—infrared imager

Demonstrations of micromachined infrared bolometers and sensors have existed for many years. But the uncooled, two-dimensional infrared imaging array from Honeywell, Inc., Minneapolis, Minnesota [23], stands out in the crowd and competes effectively with traditional designs involving cooled cameras based on group II-VI compound semiconductors (Figure 4.26).

The basic approach of the Honeywell design achieves high sensitivity to radiation by providing extreme thermal isolation for a temperature-sensitive resistive element. Incident infrared radiation heats a suspended sense resistor, producing a change in its resistance that is directly proportional to the radiation intensity. The two-level structure, consisting of an upper silicon nitride plate suspended over a substrate, provides a high degree of thermal isolation corresponding to a thermal conductance of merely 10^{-8} W/K. This value approaches the theoretical lowest limit of 10^{-9} W/K, due to radiative heat loss. The square silicon nitride plate is 50 μm on a side and 0.5-μm-thick. The thin (50 to 100 nm) resistive element rests on the silicon nitride, and has a large temperature coefficient of resistance, in the range of –0.2 to –0.3% per °C. In order to capture most or all of the incident radiation, the fill factor—the area covered by the

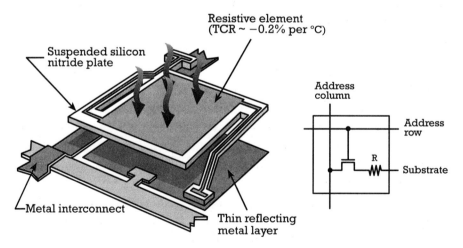

Figure 4.26 Illustration of a single sense element in the infrared imaging array from Honeywell. Incoming infrared radiation heats a sensitive resistive element suspended on a thin silicon nitride plate. Electronic circuits measure the change in resistance, and infer the radiation intensity. Adapted from Cole, et al. [23].

sensitive element as a fraction of the overall pixel area—must approach unity. The gap between the suspended plate and the substrate is approximately 1.8 μm. The silicon nitride plate and a thin reflecting metal directly underneath it form a quarter-wave resonant cavity to increase infrared absorption at wavelengths near 10 μm—corresponding to the peak radiation from a black body near 20° C. A two-dimensional array of these pixels images activity at or near room temperature, and is useful for night vision.

The basic fabrication process relies on a surface micromachining approach, but unlike the polysilicon-surface-micromachining process, it incorporates an organic layer, such as polyimide, as the sacrificial material. The fabrication of the pixels occurs after the fabrication of standard CMOS electronic circuits on the silicon substrate. In a typical array size of 240 × 336 pixels, it is nearly impossible to obtain individual leads to each element. The integrated electronics provide multiplexing as well as scan and read-out operations.

The CMOS electronic circuits are fabricated first. The last step in the CMOS process ensures that the surface is planar. One approach is by chemomechanical polishing (CMP) of a silicon dioxide passivation layer. The fabrication of the sense pixels begins with the deposition and patterning of the bottom metal films of the two-level structure. The composition of the metal does not appear to be critical. In the next step, the 1.8-μm-thick sacrificial layer is deposited. The public literature does not specify the type of material, but one could use an organic polyimide film or photoresist that can sustain the subsequent thermal cycles of the fabrication process. Standard lithography and etching methods are applied to define contacts through the sacrificial layer to the underlying metal. These contacts also serve to form anchor points for the suspended plate. A 0.5-μm-thick silicon nitride layer is deposited at low temperature and patterned, using standard lithography, in the shape of the suspended plate. The next deposition step is critical because it defines the thin, temperature-sensitive resistor. Two families of materials exhibit suitable sensing properties:

- Vanadium oxides (VO_2, V_2O_3, and V_2O_5);

- Lanthanum manganese oxides ($La_{1-x}A_x MnO_3$; A = Ca, Sr, Ba or Pb).

Sputtered vanadium oxides have a convenient sheet resistance (~ 25 kΩ per square at 25° C), acceptable $1/f$ noise, high absorption of infrared radiation, and lastly, a large temperature coefficient of resistance

(TCR) of about –0.2% per °C. Lanthanum manganese oxides yield even larger TCRs, in the range of –0.3% per °C, with low $1/f$ noise. The combination of low noise and high TCR are critical to increasing sensitivity. After the deposition and patterning of the resistive element, another silicon nitride layer is applied for encapsulation of the sensitive components. Removal of the sacrificial layer by plasma-etching releases the silicon nitride plate. Oxygen plasma is effective at isotropically removing organic materials, including polyimide and photoresist. Finally, the parts are diced, then packaged under vacuum (< 10 Pa) to reduce heat loss by conduction.

The read-out electronics apply a constant voltage pulse sequentially to each pixel, and measure the corresponding current. The estimated change in temperature for an incident radiation power of 10^{-8} W is only 0.1° C. The corresponding resistance change is a measurable –10 Ω for a 50 kΩ resistor. The thermal capacity of a pixel is 10^{-9} J/K, determined by the very small thermal mass of the suspended plate. Consequently, the thermal response time, defined by the ratio of thermal capacity to thermal conductance, is less than 10 ms, sufficiently fast for most imaging applications. The signal-to-noise ratio is limited by thermal and $1/f$ noise, to about 49 dB. Special circuits perform a calibration step that subtracts from the active image the signal of a blank scene. The latter signal incorporates the effects of non-uniform pixel resistance across the array. An intermittent shutter provides the blank scene signal, therefore allowing continuous calibration.

Carbon monoxide gas sensor

Many gas sensors operate on the principle of modulating the resistance of a metal-oxide element by adsorption of gas molecules to its surface. The adsorbed gas molecules interact with the surface of such a wide-bandgap semiconductor to trap one or more conduction electrons, effectively reducing the surface conductivity. The resistance is inversely proportional to a fractional power of the gas concentration. The class of sensor materials includes the oxides of tin (SnO_2), titanium (TiO_2), indium (In_2O_3), zinc (ZnO), tungsten (WO_3), and iron (Fe_2O_3). Each metal oxide is sensitive to different gases. For example, tin oxide is effective at detecting alcohol, hydrogen, oxygen, hydrogen sulfide, and carbon monoxide. Indium oxide, by contrast, is sensitive to ozone (O_3); zinc oxide is useful for detecting halogenated hydrocarbons. Unfortunately, most are adversely affected by humidity, which must be controlled at all times. In addition, variations in material properties require that each sensor be individually calibrated.

The MGS1100 carbon monoxide sensor from Motorola, Inc., Schaumburg, Illinois [24], incorporates a tin-oxide, thin-film sense resistor over a polysilicon resistive heater (Figure 4.27). The role of the heater is to maintain the sensor at an operating temperature between 100 and 450° C, thus reducing the deleterious effects of humidity. The sense resistor and the heater reside over a 2-μm-thick silicon membrane to minimize heat loss through the substrate. Consequently, a mere 47 mW is sufficient to maintain the membrane at 400° C. There is a total of four electrical contacts: two connect to the tin-oxide resistor, and the other two connect to the polysilicon heater. The simplest method to measure resistance is to flow a constant current through the sense element, and record the output voltage.

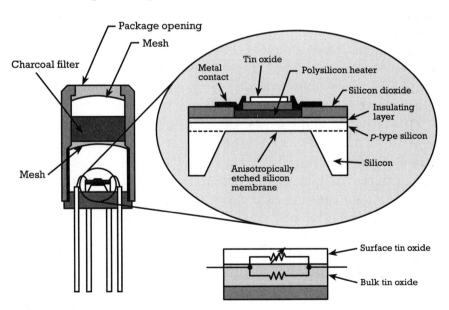

Figure 4.27 Illustration of the Motorola MGS1100 carbon monoxide sensor, its equivalent circuit model, and the final packaged part. The surface resistance of tin-oxide changes in response to carbon monoxide. A polysilicon heater maintains the sensor at a temperature between 100 and 450° C in order to reduce the adverse effects of humidity. Adapted from Lyle et al. [24].

The particulars of the fabrication process for the Motorola MGS1100 carbon monoxide sensor are not publicly disclosed, but demonstrations of similar devices exist in the literature. A simple process would begin with the forming of a heavily doped, p-type, 2-μm-thick layer of silicon either by epitaxial growth or, alternatively, by ion implantation and annealing. The deposition of a silicon nitride layer follows. A chemical vapor deposition (CVD) step provides a polysilicon film that is later patterned and etched in the shape of the heater. The polysilicon film is doped either in-situ during the CVD process, or by ion implantation and subsequent annealing. An oxide layer is then deposited, and contact holes etched in it. The purpose of this layer is to electrically isolate the polysilicon heater from the tin-oxide sense element. The tin-oxide layer is deposited by sputtering tin and oxidizing it at approximately 400° C. An alternative deposition process is sol-gel, starting with a tin-based organic precursor and curing by firing at an elevated temperature. The tin-oxide layer is patterned using standard lithography and etched in the shape of the sense element. Sputtered and patterned aluminum provides contact metallization. Finally, an etch from the back side in potassium hydroxide or ethylenediamine pyrocathechol (EDP) forms a thin membrane by stopping on the heavily doped p-type surface silicon layer. Naturally, a masking layer (e.g., silicon nitride) on the back side of the substrate and protection of the front side are necessary.

The operation of the sensor consists of applying to the heater a 5-V pulse for 5 s, followed by a 1-V pulse lasting 10 s. The corresponding temperature is 400° C during the first interval, decreasing to 80° C during the second pulse. To maintain consistency, the resistance measurement always occurs at the same time during the interval—in this case, at 9.5 s into the second 10-s long pulse. The MGS1100 sensor demonstrates a response from 1 to 4,000 ppm (parts per million) of carbon monoxide (CO), over a humidity range of 20 to 80%. The output signal shows a square-root dependence on CO concentration, with little dependence on humidity for CO concentrations above 60 ppm.

Micromachined microphone

The open literature is rich in surveys and references on micromachined microphones, each unique in its design and fabrication. Their penetration into high-volume, low-cost applications such as cellular telephony, remains hindered by the existence of inexpensive conventional microphones. Equally challenging is the precision-measurement market dominated by Brüel and Kjær A/S, Nærum, Denmark. This particular market demands strict performance standards, but is relatively small, with an annual volume of less than 100,000 units. Nevertheless, a recent demonstration from Knowles Electronics, Inc., Itasca, Illinois [25], underscores the attractiveness of MEMS technology to the hearing aid industry, where small size is of paramount importance. This micromachined microphone is described next.

A micromachined capacitive (condenser) microphone measures extremely small pressure fluctuations (less than 0.1 Pa, or 10^{-6} atm.) that arise from the transmission of acoustic energy through air. Its basic structure consists of a diaphragm suspended over a back plate, the combination of which forms a capacitor. Holes through the back plate and the supporting substrate provide a leaky path, ensuring that no static pressure builds up across the two sides of the diaphragm—otherwise, variations in atmospheric pressure are sufficient to overload the sensor's output. In operation, the diaphragm vibrates in response to dynamic acoustical pressure waves; in other words, a microphone responds to changes in pressure rather than to pressure itself. Variation in capacitance between the diaphragm and the back plate gives a direct measure of the acoustic pressure level. A back chamber (behind the back plate) forms a Helmholtz resonating cavity that tunes the acoustic impedance and the overall frequency response of the sensor (Figure 4.28).

Sensitivity, frequency response, and input-referred noise are the technical characteristics that reflect the performance of a condenser microphone. A typical condenser microphone has a sensitivity between 5 and 30 mV/Pa, and a frequency response extending from 10 Hz up to 30 kHz. The sensitivity tends to decrease at high frequencies because of air streaming through the narrow gap between the diaphragm and the back plate. If V_b is the voltage applied across the capacitor, s is the spacing between the diaphragm and the back plate, and p is the acoustic pressure, then the sensitivity is given by $(V_b/s)(ds/dp)$. A permanently stored charge in a conventional electret condenser microphone provides an equivalent voltage of several hundred volts, much higher than the 10 to 20 V typically available for a micromachined microphone. A small gap and a compliant diaphragm then become necessary to compensate, by increasing the factor $(1/s)(ds/dp)$. There are two sources of noise,

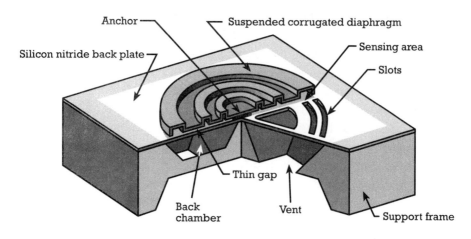

Figure 4.28 Illustration of a cut-out section of the micromachined condenser microphone from Knowles Electronics, Inc. A corrugated diaphragm vibrates in response to incoming sound. Changes in capacitance between the diaphragm and a silicon nitride back plate are measured using electronic circuits. Slots reduce the effect of air damping and ensure that no static pressure can develop across the diaphragm. Adapted from Schafer et al. [25].

mechanical and electronic. Mechanical noise is thermal in nature, and originates from the Brownian motion of the diaphragm material. Its equivalent noise pressure is equal to $(kTc)^{1/2}$, where k is Boltzmann's constant, T is temperature, and c is the acoustic impedance of the microphone. The acoustic impedance increases with decreasing volume of the back chamber. This clearly does not favor micromachined microphones for low noise applications. The thermal noise corresponding to a back chamber volume of 1 mm^3 is approximately 30 dB SPL (Sound Power Level), equivalent to background whisper.

In the Knowles microphone a corrugated circular diaphragm, anchored in its center and free to deflect at the edges, provides ample mechanical compliance. A small gap around the edge provides a controlled, pressure-relief path to equalize static pressure across the diaphragm. The high compliance of the outermost peripheral ring makes it ideal for sensing. The back plate in this annular region is perforated to reduce air damping and acoustic resistance. The smallest gap between the diaphragm and the back plate is 1 μm, increasing to 4 μm under the corrugations. The effective resting capacitance is 0.2 pF. Electronic circuits integrated on-chip supply a DC excitation voltage of 12 V, and perform sense and read-out functions.

The device fabrication integrates the electronic circuits through a standard low-voltage CMOS process, followed by additional steps to define the micromechanical structure. Immediately following CMOS fabrication, a 1.1-μm-thick silicon nitride layer is deposited by a combination of plasma-enhanced chemical vapor deposition (PECVD) and low-pressure chemical vapor deposition (LPCVD). The stress in the LPCVD film keeps it under tension. Holes corresponding to the damping slots are then formed in the silicon nitride using standard lithography and etch methods. A thin layer of chromium is sputter-deposited and patterned to define the first electrode of the capacitor. This concludes the fabrication of the back plate. The subsequent steps define the sacrificial layers and the corrugated diaphragm. A first aluminum sacrificial layer, 3-μm-thick, is deposited, then patterned and etched in the shape of the corrugation relief. A second aluminum layer, 1-μm-thick, serves as a second sacrificial layer that defines the thin gap spacing, and increases the thickness under the corrugations to 4 μm. A PECVD silicon nitride deposition step, followed by lithography and etching, define the corrugated diaphragm. The pattern design is such that the diaphragm is anchored in its center to the substrate. Sputtered chromium on the diaphragm forms the second electrode of the capacitor. The overall thickness of the diaphragm is 0.75 μm. Finally, depositions of titanium-tungsten (TiW) and gold (Au) form electrical interconnects and bond pads. The final fabrication step involves etching in potassium hydroxide. The etch proceeds through openings in a silicon nitride mask on the back side, as well as from the front side removing the aluminum sacrificial layers and releasing the silicon nitride back plate. These two anisotropic etch fronts coalesce to form four vent holes through the substrate.

Nominal sensitivity for the Knowles micromachined microphone is 10 mV/Pa, at 1 kHz. The frequency response extends from 150 Hz up to 17 kHz . The weighted input-referred noise is approximately 30 dB SPL. The noise is thermal, with equal contributions from the electronic buffer amplifier, diaphragm damping, and resistance of the pressure-relief path. These performance figures are comparable to the specifications of microphones used in broadcasting.

Actuators

The physical world is not still, but rather it is very dynamic and full of motion. If sensors extend our faculties of sight, hearing, smell, and touch, then actuators must be the extensions of our hands and fingers. They give us the agility and dexterity to manipulate physical parameters well beyond our reach. It is not surprising that the promise to control at a miniature scale is fascinating. Wouldn't the surgeon dream of electronically-controlled precision surgical tools? And what to do when our sensors tell us of a need to locally act and control on a microscopic scale? It is actuation that affords us the ability to apply this type of feedback.

The number of commercial systems or components with microactuators is limited, underscoring the nascency of this field. The following section first describes a novel display system capable of steering light on the scale of its constituent miniature mirrors, then three examples of micromachined valves. Collectively, they illustrate the current state of MEMS actuation.

Digital Micromirror Device™

The Digital Micromirror Device™—DMD™—is a trademark of Texas Instruments, Dallas, Texas, which developed and commercialized this new concept in projection display technology, referred to as Digital Light Processing™—DLP™. U.S. Patent #4,615,595 (Oct. 7, 1986) describes the early structure of the DMD™. The technology has since undergone continuous evolution and improvements. In 1996 Texas Instruments formally introduced its new product family of DLP-based projection systems.

The DMD™ consists of a two-dimensional array of optical switching elements (pixels) on a silicon substrate [26]. Each pixel consists of a reflective micromirror supported from a central post (Figure 4.29). This post is mounted on a lower metal platform—the yoke—itself suspended by thin and compliant torsional hinges from two stationary posts anchored directly to the substrate. Two electrodes positioned underneath the yoke provide electrostatic actuation. A 24-V bias voltage between one of the electrodes and the yoke tilts the mirror towards that electrode. The nonlinear electrostatic and restoring mechanical forces make it impossible to accurately control the tilt angle. Instead, the yoke snaps into a fully deflected position, touching a landing-site biased at the same

The Gearbox: Commercial MEM Structures and Systems

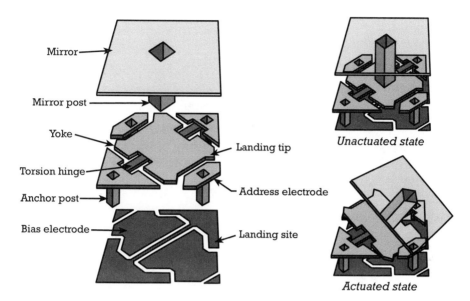

Figure 4.29 Illustration of a single DMD™ pixel in its resting and actuated states. The basic structure consists of a bottom aluminum layer containing electrodes, a middle aluminum layer containing a yoke suspended by two torsional hinges, and a top reflective aluminum mirror. An applied electrostatic voltage on a bias-electrode deflects the yoke and the mirror towards that electrode. A pixel measures approximately 17 μm on a side. Adapted from Van Kessel et al. [26].

potential—to prevent electrical shorting. The angle of tilt is limited by geometry to ±10° (the direction of the sign is defined by the optics). The restoring torque of the hinges returns the micromirror to its initial state once the applied voltage is removed. CMOS static-random-access-memory (SRAM) cells, fabricated underneath the micromirror array, control the individual actuation states of each pixel and their duration. The OFF state of the memory cell tilts the mirror by −10°, whereas the ON state tilts it by +10°. In the ON state, off-axis illumination reflects from the micromirror into the pupil of the projection lens, causing this particular pixel to appear bright. In the other two tilt states, 0° and −10°, an aperture blocks the reflected light, giving the pixel a dark appearance (Figure 4.30). This beam-steering approach provides high contrast between the bright and dark states. Each micromirror is 16 μm square, and is made of aluminum for high reflectivity. The pixels are arrayed in

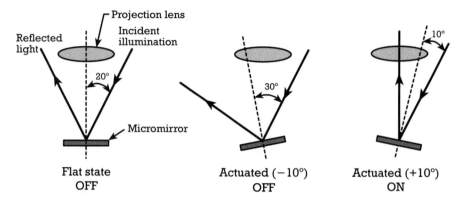

Figure 4.30 Illustration of optical beam-steering using the switching of micromirrors. Off-axis illumination reflects into the pupil of the projection lens only when the micromirror is tilted in its +10°-state, giving the pixel a bright appearance. In the other two states, the pixel appears dark [26].

two dimensions on a pitch of 17 μm to form displays, with standard resolutions from 800 × 600 pixels (SVGA) up to 1280 × 1024 pixels (SXGA). The fill factor, defined as the ratio of reflective area to total area, is approximately 90% allowing a seamless (continuous) projected image free of pixelation.

While the operation of each mirror is "only digital," in other words, the pixel is either bright or dark, the system is capable of achieving gray shades by adjusting the dwell time of each pixel—the duration is bright or dark. The mechanical switching time, including settling time, is approximately 16 μs, much faster than the response of the human eye. At these speeds, the eye can only interpret the amount—not the duration—of light it receives in a pulse. This, in effect, is equivalent to the impulse response of the eye. Modulating the duration of the pulse, or the dwell time, gives the eye the sensation of gray by varying the integrated intensity. Since the pixel switching speed is approximately 1,000-times faster than the eye's response time, it is theoretically possible to fit up to about 1,000 gray levels, equivalent to 10 bits of color depth. In actuality, full-color projection uses three DMD™ chips, one for each primary color (red, green, and blue), with each chip accommodating 8-bit color depth, for a total of 16 million discrete colors. Alternatively, by using filters on a color wheel, the three primary colors can be switched and projected using a single DMD™ chip.

The Gearbox: Commercial MEM Structures and Systems

Texas Instruments uses surface micromachining to fabricate the DMD™ on wafers incorporating CMOS electronic address and control circuitry (Figure 4.31). The basics of the fabrication process are in some respects similar to other surface-micromachining processes: the etching of one or more sacrificial layers releases the mechanical structures. But they differ in that they must address the reliable integration of close to one million micromechanical structures with CMOS electronics. All micromachining steps occur at temperatures below 400° C, sufficiently

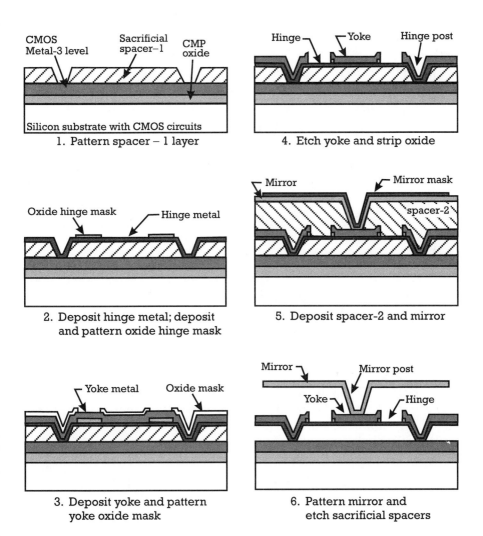

Figure 4.31 Fabrication steps of the Texas Instruments DMD™ [26].

low to ensure the integrity of the underlying electronic circuits. Standard 0.8 µm, double-metal level, CMOS technology is used to fabricate control circuits and SRAM memory cells. A thick silicon dioxide layer is deposited over the second CMOS metal layer. Chemomechanical polishing (CMP) of this silicon dioxide layer provides a flat starting surface for the subsequent building of the DMD™ structures. A third, aluminum-metal layer is sputter-deposited and patterned to provide bias and address electrodes, landing pads, and electrical interconnects to the underlying electronics. Photoresist is spin-deposited, exposed, developed, and hardened with ultraviolet (UV) light to form the first sacrificial layer. A sputter-deposition of an aluminum alloy defines the hinge metal layer. The mechanical integrity of the DMD™ relies on low stresses in the hinge. Naturally, the exact composition of the alloy remains proprietary to Texas Instruments. A thin, silicon dioxide mask is then deposited with PECVD, and patterned to protect the torsion hinge regions. The aluminum is not etched after this step. Retaining this silicon dioxide mask, another sputtering step deposits a thicker, yoke-metal layer, also made of a proprietary aluminum alloy. A thin layer of silicon dioxide is subsequently deposited and patterned in the shape of the yoke and anchor posts. An etch step removes the exposed aluminum areas down to the organic sacrificial layer. But in the regions where the oxide hinge mask remains, only the thick yoke metal is removed, stopping on the silicon dioxide mask and leaving intact the thin, torsional hinges. Both silicon dioxide masking layers are stripped before a second sacrificial layer, also made of UV-hardened photoresist, is deposited and patterned. Yet another aluminum alloy sputter-deposition defines the mirror material and the mirror post. A silicon dioxide mask protects the mirror regions during etch of the aluminum alloy.

The remaining fabrication steps address the preparation for sawing and packaging, made difficult by the delicate micromechanical structures. A wafer saw cuts the silicon along edge scribe lines to a depth that allows breaking the individual dice apart at a later stage. An oxygen-plasma etch step removes both sacrificial layers and releases the micromirrors. A special passivation step deposits a thin, antistiction layer to prevent any adhesion between the yoke and the landing pads. Finally, a singulation process breaks apart and separates the individual dice. The packaging of the DMD™ is discussed in Chapter 6.

The Gearbox: Commercial MEM Structures and Systems 147

Reliability is the *sine qua non* of the commercial success of DMD™ technology. The designs described above are the result of extensive efforts at Texas Instruments aimed at understanding the long-term operation of the pixels, as well as their failure modes. The DMD™ micromirrors are sufficiently robust to withstand normal environmental and handling conditions, including 1500-G mechanical shocks, because the weight of the micromirrors is insignificant. The major failure and malfunction mechanisms are surface contamination and hinge memory. The latter is the result of "metal creep" in the hinge material, and causes the mirror to exhibit a residual tilt in the absence of actuation voltages. Advancements in the hinge metal-alloy and fabrication processes have yielded a mean time between failure (MTBF) of more than 100,000 h.

Micromachined valves

A new generation of miniature valves with electronic control would be desirable among both manufacturers and users of valves. For example, recent trends in home appliances indicate a shift towards total electronic control [27]. Electronically programmable gas stoves, currently under development, require low-cost, electronically controlled gas valves. Moreover, miniature valves are important for the control of fluid-flow functions in portable biochemical analysis systems [28].

The field of micromachined valves remains nascent and in its infancy. In order for silicon micromachined valves to gain a substantial foothold in the market, they must effectively compete with the relatively mature, traditional valve technologies. These cover a broad range of media, pressures, flow rates, and price. It is unlikely that micromachined valves will displace traditional valves; rather, they will complement them in special applications where size and electronic control are beneficial (Table 4.4).

The following sections describe three micromachined valves. Two devices from Redwood Microsystems, Inc., Menlo Park, California; and TiNi Alloy Company, San Leandro, California, illustrate the efforts of two small companies in commercializing this technology. A third micromachined valve developed for internal use at Hewlett-Packard Laboratories, Palo Alto, California, was put on display at the San Jose Tech Museum, San Jose, California, after the company decided to halt further development. All three valves operate on the principle of blocking

Table 4.4
Some Potential Applications for Silicon-Micromachined Valves

Applications
Electronic flow regulation of refrigerant for increased energy savings
Electronically programmable gas cooking stoves
Electronically programmable pressure regulators for gas cylinders
Accurate mass flow controllers for high purity gas delivery systems
Accurate drug delivery systems
Control of fluid flow in portable biochemical analysis systems
Portable gas chromatography systems
Proportional control for electro-hydraulic braking (EHB) systems

a vertical fluid port with a silicon plug suspended from a spring that is sufficiently compliant to allow vertical displacement during actuation. Accordingly, the inlet pressure limit is low, typically less than 150 psig[2]. Otherwise, higher pressures would require a stiffer suspension spring, which in turn would necessitate a higher output force from the miniature actuator—a difficult task to accomplish on this scale.

Micromachined valve from Redwood Microsystems

Early development of this valve took place in the middle 1980s at Stanford University [29]. Redwood Microsystems was founded shortly thereafter, with the objective of commercializing the valve. The actuation mechanism of either normally-open or normally-closed valves[3] depends on the electrical heating of a control liquid sealed inside a cavity. When the temperature of the liquid rises, its pressure increases, thus exerting a force on a thin diaphragm wall and flexing it outward. In a normally-open valve, the diaphragm itself occludes a fluid port by its flexing action, hence blocking flow (Figure 4.32). Upon removal of electrical power, the control liquid entrapped in the sealed cavity cools down and the diaphragm returns to its flat position, consequently allowing flow through

---

2. The *psig* is a unit of differential (gauge) pressure equal to one psi (or 6.9 kPa).
3. The trademark name of the valve is the Fluistor™, short for fluid transistor, because the valve is electrically gated in a fashion similar to the electronic transistor.

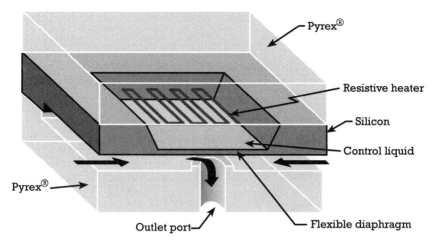

Figure 4.32 Illustration of a normally-open valve from Redwood Microsystems. Heating of a control liquid sealed inside a cavity causes a thin, silicon diaphragm to flex and block the flow through the outlet orifice. The inlet orifice is not shown.

the port. The flexing membrane is in intimate contact with the fluid flow, which increases heat loss by conduction and severely restricts the operation of the valve. A more recent demonstration from Redwood Microsystems shows a thermal-isolation scheme using a glass plate between the heated control liquid and the flexible membrane. Small perforations in the isolation glass permit the transmission of pressure to actuate the diaphragm.

The normally-closed valve uses mechanical levering activated by a liquid-filled thermopneumatic actuator to open an outlet orifice (Figure 4.33). The outward flexing action of the diaphragm, under the effect of internal pressure, develops a torque about a silicon fulcrum. Consequently, the upper portion of the valve containing the actuation element lifts the valve plug above the valve seat, permitting flow through the orifice.

The pressure that develops inside the sealed cavity results from the heating of the control liquid, which must meet some criteria in order to yield efficient actuation. In particular, the control liquid must be inert and noncorrosive. It must be electrically insulating but thermally conductive, and must boil or expand considerably when heated. Redwood Microsystems uses one of the Fluorinert™ perfluorocarbon liquids from

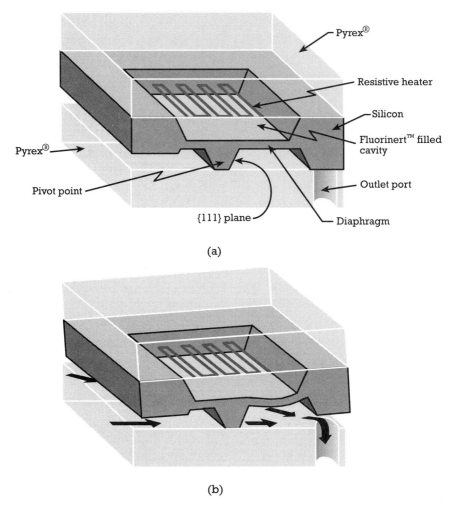

Figure 4.33 Illustration of the basic operating mechanism of a normally-closed micromachined valve from Redwood Microsystems. (a) The upper stage of the valve normally blocks fluid flow through the outlet orifice. The inlet orifice is not shown. (b) Heating of the Fluorinet™ liquid sealed inside a cavity flexes a thin, silicon diaphragm, which in turn causes a mechanical lever to lift the valve plug. Adapted from the Fluistor™ valve specification sheet (Redwood Microsystems, Menlo Park, California).

3M Chemicals, St. Paul, Minnesota. Their boiling point ranges from 56 to 250° C, and they exhibit large temperature coefficients of expansion (~ 0.13% per ° C). They are also electrically insulating and have a high dielectric constant. Clearly, the choice of control liquid determines the actuation temperature, and correspondingly, the power consumption and switching times of the valve.

The NO-1500 Fluistor™ normally-open gas valve provides proportional control of the flow rate for noncorrosive gases. The flow rate ranges from 0.1 sccm up to 1,500 sccm. The maximum inlet supply pressure is 690 kPa (100 psig)[4], the switching time is typically 0.5 s, and the corresponding average power consumption is 500 mW. The NC-1500 Fluistor™ is a normally-closed gas valve with similar pressure and flow ratings, but its switching response is 1 s and it consumes 1 W. Because the Fluistor™ relies on the absolute temperature—rather than a differential temperature—of the control liquid for actuation, the valve cannot operate at elevated ambient temperatures. Consequently, the Fluistor™ is rated for operation between 0 to 55° C. The normally-closed valve measures approximately 6 mm × 6 mm × 2 mm, and is packaged inside a TO-8 can with two attached tubes. The packaging is further discussed in Chapter 6.

U.S. Patent #4,966,646 (Oct. 30, 1990) describes the basic fabrication steps for a normally-open valve, however, the fabrication details of a normally-closed valve are not publicly available. The following process delineates the general steps to fabricate a normally-closed valve. The features in the intermediate silicon layer are fabricated by etching both sides of the wafer in potassium hydroxide. The front-side etch forms the cavity that will later fill with the actuation liquid. The etch on the bottom side forms the fulcrum, as well as the valve plug. Accurate timing of both etches ensures the formation of the thin diaphragm in the middle of the silicon wafer. The top glass wafer is processed separately to form a sputtered, thin-film metal heater. Ultrasonic drilling opens a fill hole through the top Pyrex® glass substrate, as well as the inlet and outlet orifices in the lower Pyrex® glass substrate. Both glass substrates are sequentially bonded to the silicon wafer using anodic bonding. In the final step, the Fluorinet™ liquid fills the cavity. Special silicone compounds dispensed over the fill-hole permanently seal the Fluorinert™ inside the cavity.

---

4. Fluid flow through an ideal orifice depends on the differential pressure across it. The flow is equal to $C_d A_0 \sqrt{2\Delta P/\rho}$ where ΔP is the difference in pressure, ρ is the density of the fluid, A_0 is the orifice area, and C_d is the discharge coefficient, a constant that varies from 0.95 to 0.99 depending on the geometry of the orifice.

Micromachined valve from TiNi Alloy Company

TiNi Alloy Company, San Leandro, California, is another small company with the objective of commercializing micromachined valves. Its design approach, however, is very different from that of Redwood Microsystems. The actuation mechanism relies on titanium-nickel (TiNi) [30], a shape-memory alloy, and hence the name of the company. The rationale is that shape-memory alloys are very efficient actuators and can produce a large volumetric energy density, approximately 5 to 10 times higher than competing actuation methods. It is, however, the integration of TiNi processing with mainstream silicon manufacturing that remains to be an important hurdle.

The complete valve assembly consists of three silicon wafers and one beryllium-copper spring to maintain a closing force on the valve poppet (plug) (Figure 4.34). One silicon wafer incorporates an orifice. A second wafer is simply a spacer defining the stroke of the poppet as it actuates.

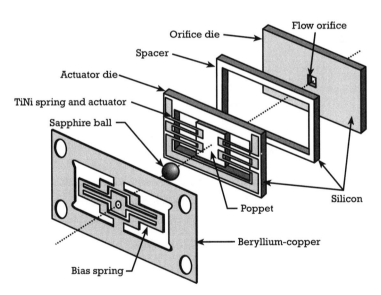

Figure 4.34 Assembly of the micromachined, normally-closed valve from TiNi Alloy Company. The beryllium-copper spring pushes a sapphire ball against the silicon poppet to close the flow orifice. Resistive heating of the TiNi spring above its transition temperature causes it to recover its original flat (undeflected) shape. The actuation pulls the poppet away from the orifice, hence permitting fluid flow. Adapted from A. D. Johnson, TiNi Alloy Company, San Leandro, California.

The Gearbox: Commercial MEM Structures and Systems

A third silicon wafer contains the valve poppet suspended from a spring structure made of a thin-film—titanium-nickel alloy. A sapphire ball between a beryllium-copper spring and the third silicon wafer pushes the poppet out of the plane of the third wafer, through the spacer of the second wafer, to close the orifice in the first wafer. Current flow through the titanium-nickel alloy heats the spring above its transition temperature (~100° C), causing it to contract and recover its original, undeflected position in the plane of the third wafer. This action pulls the poppet back from the orifice, hence permitting fluid flow.

The fabrication process relies on thin-film deposition and anisotropic etching to form the silicon elements of the valve (Figure 4.35). The fabrication of the orifice and the spacer wafers is simple, involving one etch step for each. The third wafer containing the poppet and the titanium-nickel spring involves a few additional steps. Silicon dioxide is first deposited or grown on both sides of the wafer. The layer on the back side of the wafer is patterned. A timed, anisotropic silicon etch using the silicon dioxide as a mask defines a silicon membrane. Tetramethyl ammonium hydroxide (TMAH) is a suitable etch solution because of its extreme

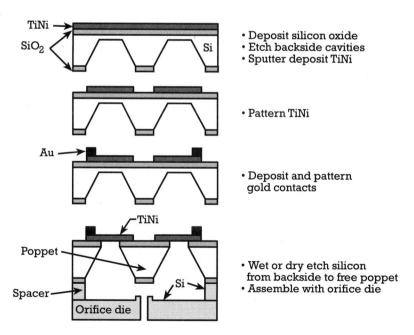

Figure 4.35 Fabrication sequence of the micromachined valve from TiNi Alloy Company. Adapted from Johnson and Shahoian [30].

selectivity to silicon dioxide. A titanium-nickel film, a few micrometers in thickness, is sputter-deposited on the front side and subsequently patterned. Double-sided lithography is critical to ensure that the titanium-nickel pattern aligns properly with the cavities etched on the back side. Gold evaporation and patterning follows; gold defines the bond pads and the metal contacts to the titanium-alloy actuator. A wet or plasma etch step from the back side removes the thin, silicon membrane and frees the poppet. At this point, the three silicon wafers are bonded together using a glass thermocompression bond. Silicon-fusion bonding is not practical because the titanium-nickel alloy rapidly oxidizes at temperatures above 300° C. Assembly of the valve elements remains manual, resulting in high production costs. The list price for one valve is about $190. Achieving wafer-level assembly is crucial in order to benefit from the cost advantages of volume manufacturing.

The performance advantage of shape-memory alloys manifests itself in low power consumption and fast switching speeds. The valve consumes less than 200 mW, switches on in about 10 ms, and off in about 15 ms. The maximum gas-flow rate and inlet pressure are 1,000 sccm and 690 kPa (100 psig), respectively. The valve measures 8 mm × 5 mm × 2 mm, and is assembled inside a plastic package.

Micromachined valve from Hewlett-Packard Laboratories

The actuation mechanism of this valve relies on the differential expansion of two heated materials. The abstract of U.S. Patent #5,058,856 (Oct. 22, 1991) describes this silicon microvalve (Figure 4.36):

> A microminiature valve having radially spaced, layered spider legs, with each leg having first and second layers of materials having substantially different coefficients of thermal expansion. The legs include heating elements and are fixed at one end to allow radial compliance as selected heating of the legs causes flexure. Below the legs is a semiconductor substrate having a flow orifice aligned with a valve face. Flexure of the legs displaces the valve face relative to the flow orifice, thereby controlling fluid flow through the orifice.

The leaf-shaped bimetallic actuator (with radial spider legs) consists of a first layer of nickel, 30-μm-thick, over a silicon membrane (the second layer). These two layers have substantially different coefficients

The Gearbox: Commercial MEM Structures and Systems

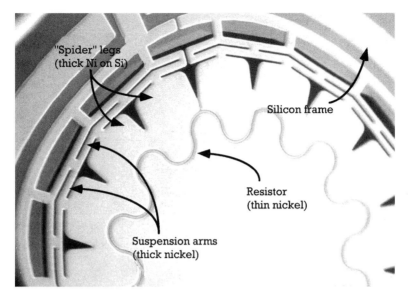

Figure 4.36 Top view photograph of the valve from Hewlett-Packard illustrating the leaf-like structure, the suspension bars and the thick- and thin-nickel regions. Courtesy of P. Barth, Hewlett-Packard, Palo Alto, California.

of thermal expansion: 13.5×10^{-6} per °C for nickel, and 2.6×10^{-6} per °C for silicon. The silicon valve face (or valve poppet) attaches to the center of the silicon membrane. Suspension arms around the periphery support the radial flexure legs and provide thermal isolation between the actuator and the silicon frame. The heater consists of a 1-μm-thick nickel serpentine element. Heating of the actuator by about 100° C above ambient causes the radial legs to flex upwards, lifting the valve plug and allowing the normally closed valve to open fully. The valve can open against a pressure of 690 kPa (100 psig) and permit a flow of 1,000 sccm of air. The power consumption is approximately 1 W, and the operating temperature range is 0 to 55° C (Figure 4.37).

The fabrication details of the valve are not publicly available. Again, one can delineate the basic steps. An upper and a lower wafer are processed separately, then bonded together at the end. A relatively shallow etch from the front side of the lower wafer defines the valve seat. Subsequent anisotropic etching from the back side in potassium hydroxide forms a square flow orifice, about 180 μm on a side. Clearly, double-sided alignment between the valve seat and the flow orifice is essential.

156   An Introduction to Microelectromechanical Systems Engineering

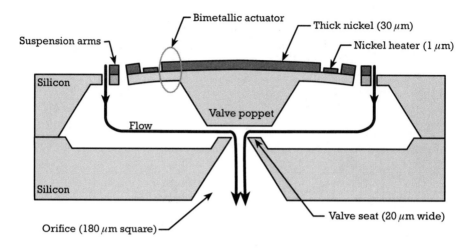

Figure 4.37 Basic cross-sectional illustration of the gas valve from Hewlett-Packard. Heating of the silicon-nickel bimetallic actuator causes it to flex upwards, lifting the poppet and opening the normally closed valve.

Sputtering of a thin (~ 1 μm) nickel layer on the front surface of the upper wafer, followed by standard lithography, defines the heater element. Patterning of thick photoresist and subsequent electroplating form the thick (~ 30 μm) nickel regions. Masked anisotropic etching of the back side of the upper wafer in potassium hydroxide forms the valve plug and silicon membrane. A dry-etch step from the front side releases the suspension bars and the spider legs. In the final step, the two wafers are bonded together.

Summary

This chapter presented a set of representative MEM structures and systems, including a number of micromachined sensors, actuators, and a few passive devices. The basic sensing and actuation methods vary considerably from one design to another, with significant consequences to the control electronics. Design considerations are many; they include the specifications of the end application, functionality, process feasibility, and economic justification.

References

[1] Timoshenko, S., "Analysis of Bi-Metal Thermostats," *Journal of the Optical Society of America*, Vol. 11, 1925, p. 233.

[2] Klaassen, E. H., et al., "Silicon Fusion Bonding and Deep Reactive Ion Etching; A New Technology for Microstructures," *Proc. 8th Int. Conf. on Solid-State Sensors and Actuators*, Stockholm, Sweden, June 25–29, 1995, pp. 556–559.

[3] U.S. Patents #3,921,916 (Nov. 25, 1975) and #3,949,410 (Apr. 6, 1976).

[4] Kneisel, L. L., J. D. Baker, and L. N. Goenka, "Silicon Micromachined CO_2 Cleaning Nozzle and Method," U.S. Patent #5,545,073 (Aug. 13, 1996).

[5] Beatty, C. C., "A Chronology of Thermal Ink-Jet Structures," *Tech. Digest Solid-State Sensor and Actuator Workshop*, Hilton Head Island, SC, June 3–6, 1996, pp. 200–204.

[6] Czarnocki, W. S., and J. P. Schuster, "The Evolution of Automotive Pressure Sensors," *Sensors*, Vol. 16, No. 5, May 1999, pp. 52–65.

[7] NPC-107 data sheet, Lucas NovaSensor, 1055 Mission Court, Fremont, California 94539, http://www.novasensor.com.

[8] Johnson, R. G., and R. E. Higashi, "A Highly Sensitive Silicon Chip Microtransducer for Air Flow and Differential Pressure Sensing Applications," *Sensors and Actuators*, Vol. 11, 1987, pp. 63–72.

[9] Suminto, J. T., "A Wide Frequency Range, Rugged Silicon Micro Accelerometer with Overrange Stops," *Proc. 9th Annual Int. Workshop on Micro Electro Mechanical Systems*, San Diego, CA, Feb. 11–15, 1996, pp. 180–185.

[10] Sasayama, T., et al., "Highly Reliable Silicon Micro-Machined Physical Sensors in Mass Production," *Proc. 8th Int. Conf. on Solid-State Sensors and Actuators*, Stockholm, Sweden, June 25–29, 1995, pp. 687–690.

[11] Chau, K. H. -L., et al., "An Integrated Force-Balanced Capacitive Accelerometer for Low-G Applications," *Proc. 8th Int. Conf. on Solid-State Sensors and Actuators*, Stockholm, Sweden, June 25–29, 1995, pp. 593–596.

[12] Offenberg, M., et al., "Novel Process for a Monolithic Integrated Accelerometer," *Proc. 8th Int. Conf. on Solid-State Sensors and Actuators*, Stockholm, Sweden, June 25–29, 1995, pp. 589–592.

[13] Van Drieënhuizen, et al., "Force-Balanced Accelerometer with mG Resolution, Fabricated Using Silicon Fusion Bonding and Deep Reactive Ion Etching," *Proc. 1997 Int. Conf. on Solid-State Sensors and Actuators*, Chicago, IL, June 16–19, 1997, Vol. 2, pp. 1229–1230.

[14] "Inertial Technology for the Future," R. R. Ragan (ed.), *IEEE Transactions on Aerospace and Electronic Systems*, Vol. AES-20, No. 4, July 1984, pp. 414–444.

[15] Emiliani, C., *The Scientific Companion*, 2nd ed., New York, NY: Wiley, 1995, pp. 204–205.

[16] Beer, F. P., and E. R. Johnston, Jr., *Vector Mechanics for Engineers: Dynamics*, 3rd ed., New York, NY: McGraw-Hill, 1977, pp. 716–719.

[17] Söderkvist, J., "Micromachined Gyroscopes," *Sensors and Actuators*, Vol. A43, 1994, pp. 65–71.

[18] Yazdi, N., F. Ayazi, and K. Najafi, "Micromachined Inertial Sensors," in Integrated Sensors, Microactuators, & Microsystems (MEMS), pp. 1640–1659, K. D. Wise (ed.), *Proceedings of the IEEE*, Vol. 86, No. 8, Aug. 1998.

[19] Chang, S., Chia, et al., "An Electroformed CMOS Integrated Angular Rate Sensor," *Sensors and Actuators*, Vol. A66, 1998, pp. 138–143.

[20] Langdon, R. M., "The Vibrating Cylinder Gyroscope," *The Marconi Review*, Fourth Quarter, 1982, pp. 231–249.

[21] Voss, R., et al., "Silicon Angular Rate Sensor for Automotive Applications with Piezoelectric Drive and Piezoresistive Read-out," *Proc. 1997 Int. Conf. on Solid-State Sensors and Actuators*, Chicago, IL, June 16–19, 1997, Vol. 2, pp. 879–882.

[22] Lutz, M. Golderer, et al., "A Precision Yaw Rate Sensor in Silicon Micromachining," *Proc. 1997 Int. Conf. on Solid-State Sensors and Actuators*, Chicago, IL, June 16–19, 1997, Vol. 2, pp. 847–850.

[23] Cole, B. E., R. E. Higashi, and R. A. Wood, "Monolithic Two-Dimensional Arrays of Micromachined Microstructures for Infrared Applications," in *Integrated Sensors, Microactuators, & Microsystems (MEMS)*, pp. 1679–1686, K. D. Wise (ed.), Proceedings of the IEEE, Vol. 86, No. 8, Aug. 1998.

[24] Lyle, R. P. and D. Walters, "Commercialization of Silicon-Based Gas Sensors," *Proc. 1997 Int. Conf. on Solid-State Sensors and Actuators*, Chicago, IL, June 16–19, 1997, Vol. 2, pp. 975–978.

[25] Schafer, D., S. Shoaf, and P. Loeppert., "Micromachined Condenser Microphone for Hearing Aid Use," *Tech. Digest Solid-State Sensor and Actuator Workshop*, Hilton Head Island, SC, June 8–11, 1998, pp. 27–30.

[26] Van Kessel, P. F., et al., "A MEMS-Based Projection Display," in *Integrated Sensors, Microactuators, & Microsystems (MEMS)*, pp. 1687–1704, K. D. Wise (ed.), Proceedings of the IEEE, Vol. 86, No. 8, Aug. 1998.

[27] "Under-Glass Controls: Creative Cooktop Delivers High Power and Fast Boiling," *Appliance Manufacturer*, July 1996, pp. 61–63.

[28] Anderson, R. C., G. J. Bogdan, A. Puski, and X. Su, "Genetic Analysis Systems: Improvements and Methods," *Tech. Digest Solid-State Sensor and Actuator Workshop*, Hilton Head Island, SC, June 8–11, 1998, pp. 7–10.

[29] U.S. Patents #4,824,073 (Apr. 25, 1989) and #4,966,646 (Oct. 30, 1990).

[30] Johnson, A. D., and E. J. Shahoian, "Recent Progress in Thin Film Shape Memory Microactuators," *Proc. IEEE Micro Electro Mechanical Systems*, Amsterdam, the Netherlands, Jan. 29–Feb. 2, 1995, pp. 216–219.

Selected bibliography

Frank, R., *Understanding Smart Sensors*, Norwood, MA: Artech House, 1996.

Kovacs, G. T. A., *Micromachined Transducers Sourcebook*, New York, NY: McGraw-Hill, 1998.

MacDonald, L. W. and A. C. Lowe (Eds.), *Display Systems: Design and Applications*, West Sussex, England: J. Wiley, 1997.

Micromechanics and MEMS: Classic and Seminal Papers to 1990, W. Trimmer (ed.), New York, NY: IEEE, 1997.

Soloman, S., *Sensors Handbook*, New York, NY: McGraw-Hill, 1998.

Wise, K. D., Editor, "Special Issue on Integrated Sensors, Microactuators, and Microsystems (MEMS)," *Proceeding of the IEEE*, Vol. 86, No.8, Aug. 1998.

CHAPTER 5

Contents

Passive micromechanical structures

Sensors and analysis systems

Actuators and actuated systems

Summary

The New Gearbox: A Peek Into the Future

I never worry about the future. It comes soon enough.

Albert Einstein, Aphorism, 1945–1946; Einstein Archive 36–570.

The great promise of MEMS technology lies in its potential to enable a new range of applications. This chapter provides a glimpse into the role of MEMS in a number of emerging applications. Naturally, significant research and development efforts are still underway, and many technical and economic questions remain to be resolved. The intent here is to provide the reader a sense of the breadth and enabling potential of the technology, but without neglecting the challenges that still prevent widespread success and market penetration. The devices and systems described next cover a number of promising applications currently under development at industrial companies, organizations, and academic institutions. These are very diverse applications ranging from genetic and chemical analysis to telecommunications. In each application, MEMS plays a critical role in enabling operation and functionality.

Passive micromechanical structures

Hinge mechanisms

Hinges are very useful passive elements in our daily lives. At the microscopic scale, they extend the utility of the inherently two-dimensional surface-micromachining technology into the third dimension (Figure 5.1). The hinge fabrication occurs simultaneously with the rest of the planar structures. Folding the hinge out of the plane gives structures access to the space above the silicon die. One potential future commercial application that may benefit from these fold-up mechanisms is the assembly of microlenses, mirrors, and other components on optical microbenches [1, 2] (Figure 5.2).

The structure is simple, consisting of a plate and a support arm made of a first polysilicon layer. A staple made of a second polysilicon layer captures the plate support arm. The staple is anchored directly to the substrate. The fabrication utilizes the polysilicon surface-micromachined process introduced in Chapter 3. The polysilicon layers are typically 2-μm-thick. The sacrificial phosphosilicate glass (PSG) layer is 0.5- to 2.5-μm-thick. Etching in hydrofluoric acid removes the PSG layer and

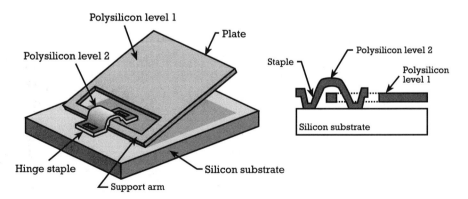

Figure 5.1 Illustration of the fold-up surface-micromachined hinge. The structure is fabricated using polysilicon surface micromachining. Adapted from Pister et al. [3].

The New Gearbox: A Peek Into the Future

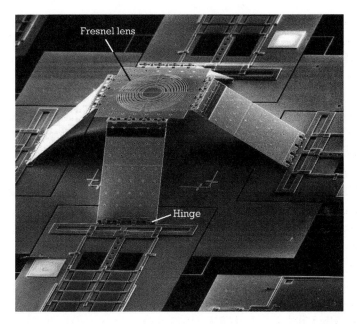

Figure 5.2 Photograph of a Fresnel microlens on an adjustable platform made of five hinged polysilicon plates. Courtesy of M. Wu, University of California, Los Angeles.

releases the mechanical plate from the substrate. Recent designs incorporate mechanical levers that snap into grooves defined in the plate and permanently lock the hinge in a vertical position.

In early demonstrations, the assembly process involved manually lifting each plate into position using sharp probes. The task remains tedious and must be automated in the future before it gains acceptance in a mainstream manufacturing environment.

Sensors and analysis systems

Miniature biochemical reaction chambers

The "medical tricorder" in the famed *Star Trek* television series is a purely fictional device for the remote scanning of biological functions in living organisms. The device remains futuristic, but significant advances in biochemistry have made it possible to decipher the genetic code of living organisms. Commercial instruments for biochemical and genetic analysis, such as those available from the Applied Biosystems Division of Perkin-Elmer, Foster City, California, perform a broad range of analytical functions, but are generally bulky. MEMS technology promises to miniaturize these instruments [4]. Witness to this potential is the high level of activity in the scientific community, and the prominence of start-up companies vying to introduce the first generation of portable biochemical diagnostics tools.

The genetic code is stored in cell chromosomes, each containing long chains of deoxyribonucleic acid (DNA). The building blocks of DNA are long molecules called nucleotides that consist of a "base" joined to a sugar-phosphate backbone. The nomenclature often interchanges between base and nucleotide to represent the same building block. There are four types of nucleotides differentiated by their bases: adenine, cytosine, guanine, and thymine. The nucleotides are labeled according to the first letter of their corresponding bases: A, C, G, and T, respectively. This is the four-letter alphabet of DNA. The sequence of nucleotides in the DNA chain contains the basic genetic information.

Each nucleotide molecule has two ends labeled 3' and 5' corresponding to the hydroxyl and phosphate groups attached to the 3' and 5' positions of carbon atoms in the backbone sugar molecule. In the long DNA chain, the 3' end of one nucleotide connects to the 5' end of the next nucleotide. This essentially gives a directionality to the DNA chain.

Two strands of DNA are joined by weak hydrogen bonds to form the well-known twisted double-helix structure. The attachment occurs between specific pairs of nucleotides; guanine bonds to cytosine (G—C), and adenine bonds to thymine (A—T). This important pairing property is known as complementarity. Color photography makes a simple analogy to understand complementarity. The three additive primary colors, red, green, and blue, are in their respective order complementary to the three subtractive colors: cyan, magenta, and yellow. A positive photographic print and its negative contain the same image information, even though the colors of the positive (the additive colors) are different from the colors of the negative (the subtractive colors). The positive and negative in photography are analogous to the two complementary strands of DNA in a double helix. Lubert Stryer's book on biochemistry [5] is recommended reading for the individual seeking detailed insight into the structure of the double helix and its chemical composition.

The New Gearbox: A Peek Into the Future

A primary objective of genetic diagnostics is to decipher the sequence of nucleotides in a DNA fragment after its extraction and purification from a cell nucleus. But the task is difficult due to the miniscule concentration of DNA available from a single cell. As a solution, scientists resort to a special biochemical process called amplification to create a large number of identical copies of a single DNA fragment. One amplification method is polymerase chain reaction (PCR). Invented in the 1980s by Kary Mullis, for which he was awarded the Nobel Prize in Chemistry in 1993, it allows the replication of a single DNA fragment using complementarity. The basic idea is to physically separate—or denature—the two strands of a double helix, then use each strand as a template to create a complementary replica.

The polymerase chain reaction begins by raising the temperature of the DNA fragment to 92° C in order to denature the two strands. Incubation occurs next at 65° C in a solution mix containing a special enzyme (called DNA polymerase, an example of which is Taq polymerase); an ample supply of nucleotides (dNTPs); and primers. The latter are short chains of nucleotides previously synthesized to hybridize—or to specifically match up using complementarity—with a very small segment of the longer DNA fragment, and consequently define the starting point for the replication process. The DNA polymerase enzyme catalyzes the reconstruction of the complementary DNA strand beginning from the position of the primer and always proceeding in the 5′ → 3′ direction. The cycle ends with two identical double helixes in addition to the starting DNA template. Repetition of the cycle geometrically increases the number of identical copies. The number of cycles is usually between 20 and 30, beyond which loss of efficiency degrades the replication process (Figure 5.3).

Over the last few years there have been several demonstrations of PCR on a silicon chip. The following section describes a silicon miniature PCR thermal cycling chamber developed at Lawrence Livermore National Laboratories, Livermore, California [7]. A version of this chamber is at the core of a portable analytical instrument under development at Cepheid, Sunnyvale, California. The micromachined chamber thermally cycles a solution between the denaturing and incubation temperatures, 92° C and 65° C, respectively. It consists of a cavity etched in a silicon wafer and sealed with a glass substrate. Grooves in the silicon—or the glass—into which disposable plastic tubes may be inserted provide access to the

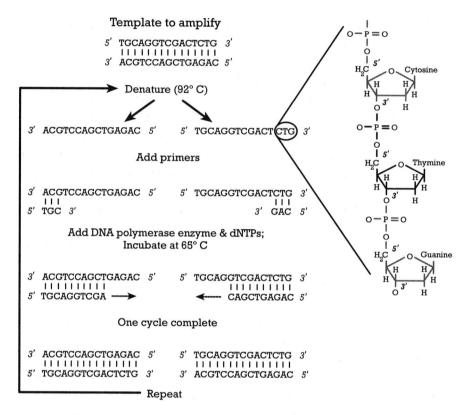

Figure 5.3 Illustration of the polymerase chain reaction. Denaturing of the starting DNA template at 92° C yields two strands, each containing all the necessary information to form a complementary replica. The addition of primers defines the starting point for replication. At 65° C, the DNA polymerase enzyme catalyzes the reconstruction of the complementary DNA strand from an ample supply of nucleotides (dNTPs). The reconstruction always proceeds in the 5' → 3' direction. The cycle ends with two identical double helixes in addition to the starting template. The cycle is then repeated. The exploded view of three nucleotides (CTG) in the denatured template shows their chemical composition, including the 3'-hydroxyl and 5'-phosphate groups. Adapted from L. Stryer [5] and Darnell et al. [6].

chamber. A silicon nitride membrane on the opposite side of the cavity supports a polysilicon heating element for thermal cycling (Figure 5.4).

The fabrication is simple, beginning with the deposition of a silicon nitride layer followed by the deposition and patterning of polysilicon. A second silicon nitride deposition encapsulates the polysilicon heater and

The New Gearbox: A Peek Into the Future

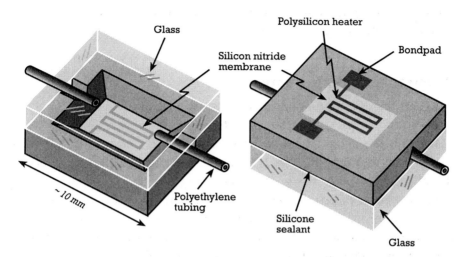

Figure 5.4 Illustrations of the front side (left) and back side (right) of a micromachined silicon polymerase chain reaction (PCR) chamber. A polysilicon heater on a silicon nitride membrane cycles the solution between the denaturing and incubation temperatures of PCR. Adapted from Northrup et al. [7].

provides a masking layer on the opposite side of the wafer. Double-sided lithography and etching are used to define openings in this masking silicon nitride layer. Anisotropic etching in potassium hydroxide opens the cavity as well as the access grooves. The etch stops on the silicon nitride film on the opposite face. The glass substrate is subsequently glued to the silicon wafer using a silicone adhesive.

The small silicon chamber typically holds about 50 μL. The small thermal mass of the chamber and the fluid makes it possible to ramp up the temperature at rates in excess of 10° C/s, compared to less than 1° C/s for conventional commercial instruments. A single PCR cycle takes less than one minute and consumes significantly less power. The small cavity size reduces the necessary volume of reagents—a savings in operational costs. Furthermore, many chambers can be arrayed to allow the simultaneous amplification of a large number of distinctly different DNA fragments. Recent results suggest that the miniaturization of PCR can lead to higher amplification factors not possible using conventional macroscopic devices. Surface interactions between the silicon, or subsequently deposited thin coatings, and the PCR chemistry remain under study. However, a key economic question remains: Can silicon bring about sufficient performance improvements to displace the existing and inexpensive polycarbonate plastic cartridges?

Electrophoresis on a chip

Determining the sequence of nucleotides in a DNA strand involves amplification and chemical labeling of the amplified DNA fragments with specific fluorescent or radioactive tags, and a subsequently distinct detection step which analyzes the labeled DNA products. The entire process is called "DNA sequencing." Its underlying principles are beyond the scope of this book, but the eager reader is again referred to Stryer's book on biochemistry [5]. One detection technique is electrophoresis, which employs the separation of polar molecules, including DNA, in colloidal suspension under the effect of an electric field. The rate of movement and degree of separation of the molecules are characteristics of their structure. In gel electrophoresis, DNA products are introduced at the edge of a porous gelatinous sheet, the size of a standard page. In capillary electrophoresis [8], the products are fed into a long, thin capillary tube. Separation under a large electric field positions the light molecules further downstream from the heavy molecules.[1] Optical imaging of a fluorescent tag on the 5′-end of each DNA product reveals its location in the gel or the capillary. Alternatively, the tags in gel electrophoresis may contain radioactive probes (^{32}p) and imaging occurs with a photographic film. The information from electrophoretic separation is sufficient for the biochemist to consequently infer the sequence of nucleotides in the DNA strand.

Miniaturization brings many benefits to capillary electrophoresis. Reducing the size of the capillary decreases the applied voltage required to maintain a high electric field, from a few kilovolts down to hundreds of volts. Faster separation times also become possible because the molecules have to travel shorter distances. Additionally, the overall volume of DNA and reagents decreases significantly to less than 10^{-9} liter. Early demonstrations of capillary electrophoresis on a chip took place in 1992 at Ciba_Geigy, Ltd., Basel, Switzerland [9], but the research activities

---

1. In actuality, the separation occurs according to the charge-mass ratio. For example, if two molecules have the same charge but different masses, the lighter one will move faster. Additionally, if two molecules have the same mass but different charges, the one with the smaller charge will move slower.

quickly spread to major laboratories of analytical chemistry throughout the world.[2]

The bulk of ongoing research activities in this field focuses on developing operations for the handling, steering, and detection of fluids, and on understanding the chemistry and effects of scaling on fluid flow. For example, one of the challenges remains in the precise injection of minute sample volumes in order to avoid smearing of signals by diffusion during imaging. The role of micromachining is largely secondary, affecting mostly the fabrication of fluid channels with small and precise cross-sections, typically measuring less than 100 μm in width.

Woolley and Mathies [10] from the University of California, Berkeley, were the first, in 1995, to demonstrate DNA-sequencing by capillary electrophoresis on a glass chip (Figure 5.5). The structure of their device consists of two orthogonal channels etched in a first glass substrate: a short channel for injecting fluid and a long channel for separating the DNA fragments. A second glass substrate covers the channels and is secured to the first substrate with an intermediate adhesive. Holes etched in the top glass substrate provide fluid access ports to the embedded channels. Both channels are 50-μm wide and 8-μm deep; the separation channel is 3.5-cm long.

The fluid containing the DNA fragments is admitted into the injection channel and electrokinetically pumped by means of an electric field of 170 V/cm applied across the two ends of the channel for a duration of 30 to 60 s. Once the injection channel is filled, the applied voltage is switched across the two ends of the separation channel. The applied electric field directs a small fluid plug at the intersection of the two channels into the separation channel. At an applied electric field of 200 V/cm, it takes approximately 13 minutes to complete the separation of the DNA fragments inside the fluid plug. This compares with 8 to 10 hours to complete an equivalent separation using conventional gel electrophoresis, or 1 to 2 hours with conventional capillary electrophoresis. Optical imaging of a fluorescent tag on each DNA fragment was used to detect the separated products inside the channel. Optical fluorescence of tags is standard in conventional DNA sequencing; each of the four different types of tags binds specifically to one of the four bases, and fluoresces at a different

---

2. The reader will find extensive coverage of the research activities in this field in past proceedings of the conference on Micro Total Analysis Systems (μTAS).

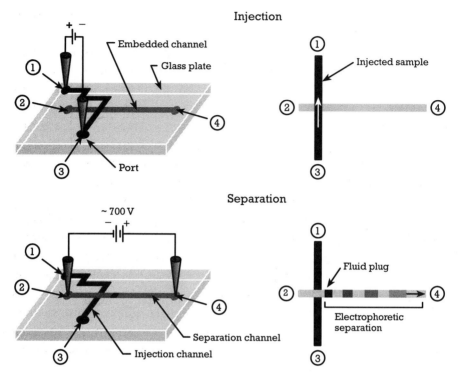

Figure 5.5 Illustration of the fluid injection and separation steps in a miniature DNA electrophoresis system. An applied electric field electrokinetically pumps the fluid molecules between ports 1 and 3 during the injection step. Another applied voltage between ports 2 and 4 initiates the electrophoretic separation of the DNA molecules. The smearing of the fluid plug in the separation channel is schematically illustrated. The capillary channels are 8 × 50 μm^2 in cross section. The separation capillary is 3.5-cm long. Adapted from Woolley and Mathies [10].

wavelength. The results from Woolley and Mathies indicate a resolution of a single nucleotide in DNA strands that are up to 500 nucleotides long.

Though the above demonstration is an important accomplishment, much remains to be done before portable DNA-sequencing instruments are available on the market. For instance, advancements in the chemistry of fluorescent tags coupled with higher sensitivity optical detectors are necessary to improve the detection resolution of narrow bands of DNA fragments, and compensate for the loss of sensitivity due to the decrease in sample volume. Furthermore, a complete sequencing system must integrate PCR with electrophoresis—or some other DNA detection method—and include all fluid preparation and handling functions such as pumping, valving, filtering, mixing of reagents, and rinsing. This demands the development of a complete system with many enabling technologies, MEMS being only one of them.

Microelectrode arrays

Electrodes are extremely useful in the sensing of biological and electrochemical potentials. In medicine, electrodes are commonly used to measure bioelectric signals generated by muscle or nerve cells. In electrochemistry, electric current from one or many electrodes can significantly alter the properties of a chemical reaction. It is natural that miniaturization of electrodes is sought in these fields, especially for applications where size is important, or arrays of electrodes can enable new scientific knowledge. Academic research on microelectrodes abounds. The reader will find a comprehensive review of microelectrodes and their properties in a book chapter by Gregory Kovacs [11].

In simple terms, the metal microelectrode is merely an intermediate element that facilitates the transfer of electrons between an electrical circuit and an ionic solution. Two competing chemical processes, oxidation and reduction, determine the equilibrium conditions at the interface between the metal and the ionic solution. Under oxidation, the electrode loses electrons to the solution; reduction is the exact opposite process. In steady-state, an equilibrium between these two reactions gives rise to an interfacial space-charge region—an area depleted of any mobile charges, electrons, or ions—separating a surface sheet of electrons in the metal electrode from a layer of positive ions in the solution. This is similar to the depletion layer at the junction of a semiconductor *p-n* diode. The interfacial space-charge region is extremely thin, measuring approximately 0.5 nm, and resulting in a large capacitance on the order of 10^{-5} F per cm^2 of electrode area. Incidentally, this is precisely the principle of operation in electrolytic capacitors. A simple electrical model for the microelectrode consists of a capacitor in series with a small resistor that reflects the resistance of the electrolyte in the vicinity.

The fabrication of microelectrode arrays first involves the deposition of an insulating layer, typically silicon dioxide, on a silicon substrate. Alternatively, an insulating glass substrate is equally suitable. A thin metal film is sputtered or evaporated, and then patterned to define the electrical interconnects and electrodes. Gold, iridium, and platinum are excellent choices for measuring biopotentials, as well as for electrochemistry. Silver is also important in electrochemistry because many published electrochemical potentials are referred to as silver/ silver-chloride electrode. It is important to note that wire bonding to platinum or iridium is very difficult. If the microelectrode must be made of such metals, then it becomes necessary to deposit an additional layer of gold over the bond pads for wire bonding. The deposition of a silicon nitride layer seals and protects the metal structures. Openings in this layer define the microelectrodes and the bond pads (Figure 5.6). The following sections describe three instances where microelectrodes show promise as diagnostics tools in the fields of biochemistry, biology, and chemistry.

DNA addressing with microelectrodes

A unique and novel application patented by Nanogen, San Diego, California [12], makes use of microelectrode arrays in the analysis of DNA fragments of unknown sequences. The approach exploits the polar property of DNA molecules to attract them to positively charged microelectrodes in an array. The analysis consists of two sequential operations, beginning first with building an array of known DNA capture probes over the electrode array, then followed by hybridization of the unknown DNA fragments. DNA capture probes are synthetic short chains of nucleotides of known specific sequence. Hybridization is the process whereby unknown DNA strands match up and bind with complementary DNA capture probes.

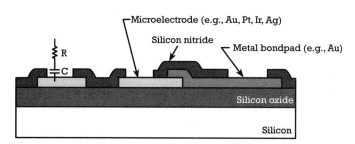

Figure 5.6 Cross-section of a microelectrode array showing two different metals for the electrodes and for the bond pads. The schematic also illustrates a basic electrical-equivalent circuit that emphasizes the capacitive behavior of a microelectrode. The silicon substrate and the silicon dioxide dielectric layer may be substituted by an insulating glass substrate.

Applying a positive voltage to a selection of microelectrodes in the array attracts DNA capture probes to these biased electrodes, where they chemically bind by means of an intermediate proprietary permeation layer. Microelectrodes in the array that are negatively biased remain clear. DNA capture probes from a first solution concentrate under the effect of the electric field, and bind to a first set of positively biased electrodes. Subsequent washing removes only unbound probes.

Immersion in a second solution binds a second type of DNA capture probes to another set of biased electrodes. Repetition of the cycle with appropriate electrode-biasing sequentially builds a large array containing tens, and potentially hundreds, of individually distinct sites of DNA capture probes, differing by their sequence of nucleotides. The removal of a capture probe from a particular site, if necessary, is simply accomplished by applying a negative potential to the desired microelectrode and releasing the probe back into the solution. It is this electrical addressing scheme to selectively attract or repel DNA molecules which makes this method versatile and powerful (Figure 5.7).

Once the array of DNA capture probes is ready, a sample solution containing DNA fragments of unknown sequence (target DNA) is introduced. These fragments hybridize with the DNA capture probes—in other words, the target DNA binds only to DNA capture probes containing a complementary sequence. Optical imaging of fluorescent tags reveals the hybridized probe sites in the array, and consequently information on the sequence of nucleotides in the target DNA. This approach is particularly beneficial in the detection of specific gene mutations, or in the search for known pathogens.

Positive biasing of select electrodes during the hybridization phase accelerates the process by actively steering and concentrating, with the applied electric field, target DNA molecules onto desired electrodes. Accelerated hybridization occurs in minutes, rather than the hours typical of passive hybridization techniques. The method is sufficiently sensitive to detect single-base differences and single-point mutations in the DNA sequence.

Cell cultures over microelectrodes

Many types of cells, in particular nerve and heart cells, can grow in an artificial culture over a microelectrode array (Figure 5.8). The growth normally requires a constant temperature at 37° C, a suitable flow of

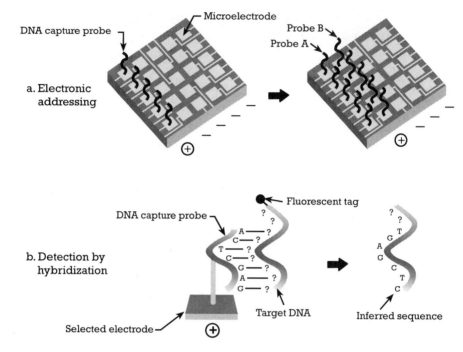

Figure 5.7 Illustration of the Nanogen electronic addressing and detection schemes. (a) A positive voltage attracts DNA capture probes to biased microelectrodes. Negatively biased electrodes remain clear of DNA. Repetition of the cycle in different solutions with appropriate electrode-biasing sequentially builds an array of individually distinct sites of DNA capture probes that differ by their sequence of nucleotides. (b) A DNA fragment with unknown sequence hybridizes with a DNA capture probe with a complementary sequence. Fluorescence microscopy reveals the hybridized site, and consequently the unknown sequence.

oxygen, and a continuous supply of nutrients [13]. Bioelectric activity, or action potential, capacitively couples across the cell membrane and the surrounding fluid to the nearest microelectrode, which then measures a small AC potential, typically between 10 and 1,000 μV in peak amplitude. The array of microelectrodes essentially images the dynamic electrical activity across a large sheet of living cells. The measured action potentials and their corresponding temporal waveforms are characteristic of the cell type and the overall health of the cell culture. For example, toxins that block the flow of sodium or potassium ions across the cell membrane suppress the action potentials or alter their frequency content [13]. This

The New Gearbox: A Peek Into the Future

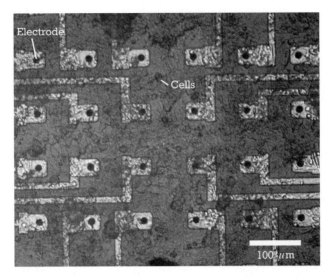

Figure 5.8 Photograph of a cultured syncytium spontaneously beating over a microelectrode array. The platinum electrodes are 10 μm in diameter with a spacing of 100 μm. The electrodes measure the extracellular currents generated by a traveling wave of action potential across the sheet of living cells. Courtesy of B. D. DeBusschere, Stanford University, Stanford, California.

approach may be useful in the future for studying the effects of experimental drugs *in vitro*, or for the early detection of airborne toxic particles.

Chemical sensing of trace metals with microelectrodes

Miniaturization of electrodes brings benefits to anodic stripping voltammetry (ASV), a well-known conventional technique in electrochemistry for the analysis of dilute samples of metals. The method employs a preconcentration step whereby electrolytic deposition collects trace metals from a large solution volume onto a mercury hemisphere previously electroplated on a metal electrode [14]. During this step, the applied potential holds the electrode at a cathodic (negative) potential with respect to the solution to attract the metal ions to the mercury hemisphere. The trace metals in the solution plate by reduction onto the surface of the mercury, and dissolve within it by amalgamation. The analysis itself involves the redissolution—or stripping—of the deposited metals back into the solution while simultaneously measuring the current. The applied voltage on the electrode is swept from zero to anodic (positive) with respect to the solution. As it reaches the oxidation potential of a metal in the mercury, the metal atoms lose electrons and redissolve into the solution. Consequently, current flows, reaches a peak, then decreases as the metal species depletes within the mercury. The measured current-voltage (I-V) curve shows multiple peaks at different voltages corresponding to the oxidation potentials of the trace metals dissolved within the mercury. The integrated electric charge under each peak is a measure of the amount of metal that was oxidized, and hence is proportional to the initial concentration of that same metal in the solution. The technique can be very sensitive and is useful to measure trace metals, for example, lead, copper, or zinc—but not mercury—in water and soil.

The measured Faradaic current is proportional to the electrode area and inversely proportional to the radius of curvature of the mercury hemisphere. It is therefore an objective to improve the signal-to-noise ratio in the measured current by decreasing the radius of curvature of the mercury ball, without reducing the effective electrode area. This is precisely the role of microelectrodes. Designing an array of microelectrodes with diameters down to 20 μm and electrically connected in parallel decreases the radius of curvature of the mercury hemisphere without reducing the total effective electrode area (Figure 5.9). The spacing between the electrodes must be sufficiently large (> 100 μm) so that the diffusion layers remain spherical and do not overlap. Researchers at Stanford University [15] demonstrated, using such microelectrodes, a detection resolution of 1 ppb (part per billion) of cadmium, lead, copper, and zinc in water. Chemtrace Corporation, Hayward, California, applied the same technique to measure traces of copper and zinc in hydrofluoric acid [16]. The purpose was to detect trace contamination in process chemicals widely used in the semiconductor industry. It is a well-known fact in the integrated circuit industry that metal contaminants are detrimental to the reliability and operation of electronic devices.

Actuators and actuated systems

Micromechanical resonators

Opening the cover of a modern cellular telephone reveals a myriad of discrete passive and active components occupying substantial volume and weight. The market's continued push for small portable telephones

The New Gearbox: A Peek Into the Future   177

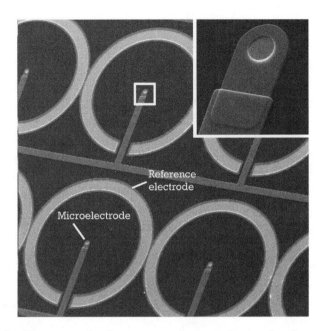

Figure 5.9 Photograph of an array of microelectrodes for the detection of traces of heavy metals using anodic stripping voltammetry. Mercury hemispheres are later electroplated on each microelectrode. The electrodes are electrically in-parallel to provide a large effective conduction area. Plated mercury hemispheres have very small radii of curvature, defined by the diameter of the microelectrode. A small radius of curvature improves the signal-to-noise ratio in anodic stripping voltammetry, and consequently, the detection resolution of trace metals. The inset shows a magnified view of an individual microelectrode. Courtesy of A. Flannery, Stanford University, Stanford, California.

argues a convincing economic case for the miniaturization of components. MEMS technology promises to deliver miniature integrated solutions, including oscillators, filters, switches, and tunable capacitors, to potentially replace conventional discrete components such as quartz crystals. However, while the potential of MEMS is exciting, one ought not to forget that meeting the aggressive price requirements dictated by the competitive nature of the telecommunications markets will be the ultimate factor in determining the level of penetration MEMS technology can achieve in wireless systems.

Quartz crystals and inductors remain at the core of every electrical resonant circuit and filter because integrated electronic oscillators cannot achieve the large quality factors (Q) necessary for the stable operation of frequency-selective communications systems. For example, a bandpass filter consisting of a network of inductors, capacitors, and resistors (an RLC ladder), with a center frequency of 70 MHz and a nominal bandwidth of 200 kHz, suffers an insertion loss of about 1 dB if the Q of the RLC circuit is 10,000—in other words, the signal suffers an undesirable attenuation of 11%. The insertion loss rapidly increases to 20 dB for a Q of 1,000—that is an attenuation by a factor of 10! If micromechanical resonators can demonstrate high Q over a wide range of tunable frequencies, then integrating them with electronics will consequently lead to system miniaturization. The frequencies of interest cover the range between 800 MHz and 2.5 GHz for front-end wireless reception, as well as the intermediate frequencies[3] (IF) at 455 kHz and above.

Basic physics tell us that a mechanical system consisting of a mass, M, and spring constant, k, resonates at a natural frequency $\frac{1}{2\pi}\sqrt{k/M}$. It follows immediately that a reduction in size brings about a decrease in mass and stiffening of the spring, thereby increasing the resonant frequency. This is the basic argument for the micromachining of resonators. The various designs differ in their implementation of excitation and sense mechanisms. The most common resonator design uses electrodes for electrostatic excitation and capacitive sensing[4] of resonance. Packaging in vacuum eliminates viscous air damping in order to obtain high Q. Scientists at the University of California, Berkeley; and the University of Michigan, Ann Arbor [17]; have demonstrated micromachined resonators operating up to 70 MHz with Q values in excess of 20,000. Fabrication was completed using the polysilicon surface-micromachining process presented in Chapter 3. Future operation in the GHz regime is not unreasonable, with springs measuring a few micrometers in length. However, it poses a number of fabrication challenges because frequency precision will require the accurate control of critical dimensions at the scale of nanometers.

---

3. A receiver converts the frequency of a selected incoming RF signal to a fixed intermediate frequency (IF) by heterodyning the signal with the local oscillator. This allows the remaining circuits in the receiver to remain precisely tuned to the intermediate frequency regardless of the frequency of the incoming signal. The following frequencies are generally considered IF: 50 kHz, 100 kHz, 262 kHz, 455 kHz, 500 kHz, 9 MHz, 10.7 MHz, 45 MHz, and 75 MHz.

4. Chapter 4 describes in greater detail capacitive sensing and electrostatic actuation.

Electrostatic comb structures are useful mechanisms for exciting resonance and for capacitively sensing the corresponding displacement. A micromachined resonator from the University of Michigan, Ann Arbor [17], incorporates two comb structures connected by a shuttle plate, and suspended from a double folded-beam spring anchored in the center to the underlying substrate (Figure 5.10).

A voltage, V_p, applied to the folded spring and shuttle plate, provides a DC bias to both comb structures. The total excitation voltage consists of an AC drive signal, v_d, at a fundamental frequency, ω_d, superposed over the DC bias, V_p. This configuration is necessary because the attractive electrostatic force varies as the square of the excitation voltage (see Chapter 4), thereby doubling the effective excitation frequency; an offset DC bias restores the fundamental frequency as the main excitation frequency. If the DC bias is much larger than the AC amplitude, it further minimizes the effect of the second harmonic. The capacitance, C, of the sense comb structure varies in time with the oscillation, resulting in an AC output current, i_0, proportional to $V_p \mathrm{d}C/\mathrm{d}t$. A transimpedance amplifier, or simply a resistor, converts the current into an output voltage which is fed back, with an appropriate gain factor, to the actuating comb structure in order to sustain the oscillation. An example of a double-comb resonating structure having 185-μm-long, 2-μm-wide, and 2-μm-thick spring beams with an effective mass of 5.7×10^{-11} kg, oscillates in the plane at a natural frequency of 16.5 kHz. A peak AC excitation of 1 mV superposed over a 20-V DC bias is sufficient to set the device into resonance. The peak velocity of the shuttle plate is on the order of 1 m/s. The measured quality factor at a pressure of 2 Pa is 23,400. The temperature dependence of the Young's modulus of polysilicon and thermal linear expansion give the resonator a temperature coefficient-of-frequency of approximately -10^{-5} per °C, worse than for quartz. Electronic compensation of the thermal error becomes necessary for long-term stability.

An additional property of the resonator is heterodyning the main drive signal with a supplemental AC carrier signal, v_c, applied across the sense capacitor. The frequency content of the output includes the two main frequencies, ω_d and ω_c, as well as their sum and difference. The heterodyning comes about when the output current is proportional to $\mathrm{d}(CV)/\mathrm{d}t$, where V is the voltage across the sense capacitor, C. The added carrier signal is a component of V, and the sense capacitance varies with a frequency, ω_d, hence the output contains the product of both AC signals.

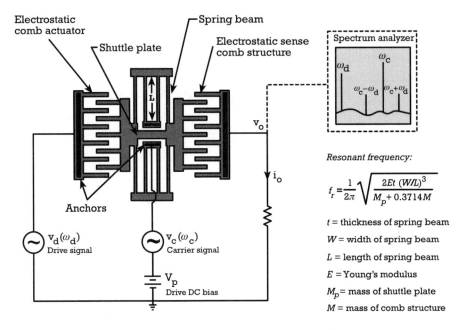

Figure 5.10 Illustration of a folded-beam comb-drive surface-micromachined resonator. One comb drive actuates the device at a frequency ω_d. A capacitive sense comb structure measures the corresponding displacement. A supplemental carrier signal at frequency ω_c can be heterodyned with the main drive signal, as shown in the frequency content of the output.

High-frequency filters

Micromechanical high-frequency bandpass filters can be built using two or more identical micromachined resonators in a linear chain, each coupled with its immediate neighbors by intermediate weak flexure springs [17]. The weak coupling between adjacent oscillators effectively creates a narrow passband of allowed frequencies, instead of a single resonant frequency.

To visualize this complex effect, let us imagine a swinging pendulum; the device can freely oscillate at its natural (resonant) frequency. Weakly coupling the mass of the first pendulum, say, with a soft rubber elastic band, to the mass of a second identical pendulum restricts the allowed oscillations of this two-body system (Figure 5.11). Now, the two masses can move either in-phase or out-of-phase with respect to each other; these are the two oscillation modes of the system. When the motions are

The New Gearbox: A Peek Into the Future

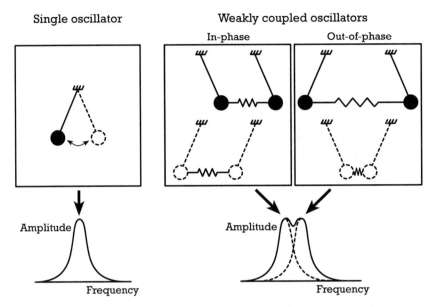

Figure 5.11 Illustration of the effect of coupled oscillators on frequency response. Two identical oscillators, weakly coupled by a spring, exhibit a bandpass frequency response. The separation between the two peaks depends on the stiffness of the spring.

in-phase, there is no relative displacement between the two masses, and consequently, no restoring force from the rubber band. The oscillation frequency of this first mode is then equal to the natural frequency of a single pendulum. But when the two masses move out-of-phase with respect to each other, their displacements are in opposite directions at any instant in time. This motion produces the largest relative displacement across the coupling rubber band, thereby resulting in a restoring force which, according to Newton's second law, provides a higher oscillation frequency. The physical coupling effectively splits the two overlapping resonant frequencies (of the two identical resonators) into two distinct frequencies, with a frequency separation dependent on the stiffness of the coupling spring. In physics, it is said that the coupling lifts the degeneracy of the oscillation modes. For a very compliant coupling spring, the two split frequencies are sufficiently close to each other that they effectively form a narrow passband. Increasing the number of coupled oscillators in a linear chain simply widens the extent of this passband. In general, the total number of oscillation modes is equal to the number of coupled oscillators in the chain.

Another explanation of the effect invokes the concept of traveling-waves. Vibrations travel as waves along a linear chain of coupled oscillators—similar to sound causing air molecules to vibrate. From the perspective of wave mechanics, the constructive and destructive interference between the different oscillation waves give rise to allowed and forbidden frequency bands—effectively, passbands. This effect is prevalent in crystal vibrations, known to physicists as phonons [18].

From the perspective of an electrical engineer, a dual electrical network accurately models the behavior of a filter made of coupled micromechanical resonators. The dual of a spring-mass system is a network of inductors and capacitors (LC network): The inductor is the dual of the mass (on the basis of kinetic energy), and the capacitor is the dual of the spring (on the basis of potential energy). A linear chain of coupled, undamped, micromechanical resonators becomes equivalent to a LC ladder network. This duality allows the implementation of filters of various types using polynomial synthesis techniques, including Butterworth and Chebyshev, common in electrical filter design. Widely available "cookbooks" of electrical filters provide appropriate polynomial coefficients and corresponding values of circuit elements [19].

A simple coupled system from the University of Michigan [17] consists of two clamped-beam oscillators with a cross-coupling flexure (Figure 5.12). Excitation occurs using a combination of a DC bias and an AC drive between the conductive clamped beams and electrodes on the substrate, 0.1 μm below. Sensing is capacitive between the beams and the electrodes. Each resonant beam is 41-μm-long, 8-μm-wide, and 2-μm-thick. The coupling flexure is 20-μm-long and 0.75-μm-wide. The passband filter has a center frequency of 7.81 MHz, with a bandwidth of 15 kHz, and an insertion loss less than 2 dB. The DC bias plays an additional role in tuning the resonant frequency of the clamped beams. The electrostatic attractive force pulls the suspended beams towards the substrate, effectively increasing their spring constant due to the stretching action. The effect is nonlinear: The resonant frequency changes by a multiplying factor equal to $\sqrt{1 - V_p C / (kd^2)}$, where d and C are the gap spacing and capacitance between the beam and electrode, respectively, without an applied bias; V_p is the applied DC bias; and k is the spring constant.

The New Gearbox: A Peek Into the Future

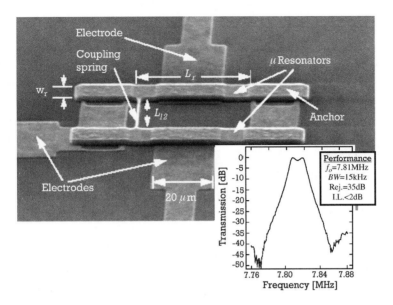

Figure 5.12 Photograph of a polysilicon surface-micromachined bandpass filter consisting of two clamped resonant beams coupled by a weak intermediate flexure spring. The excitation and sensing occur between the beams and electrodes beneath them, on the surface of the substrate. Each resonant beam is 41-μm-long, 8-μm-wide, and 2-μm-thick. The coupling flexure is 20-μm-long and 0.75-μm-wide. © 1996 IEEE [17].

"Grating light valve" display

The grating light valve, or simply GLV™, is a novel display concept invented initially at Stanford University. Silicon Light Machines, Sunnyvale, California, is developing a commercial product based on the licensed technology [20]. The fundamental light-switching concept relies on closely spaced parallel rows of reflective ribbons suspended over a substrate. The separation gap between the ribbons and the substrate is approximately one-quarter the wavelength of light in the visible. In their resting state, the ribbons appear as a continuous surface to incident light, and normal reflection occurs. But when an electrostatic voltage pulls down alternate rows of ribbons, the light reflecting from the deflected ribbons travels an additional one-half of a wavelength (twice the gap), and thus becomes 180° out-of-phase with respect to the light from the stationary ribbons. This effectively turns the ribbons into a phase grating, diffracting the incident light into higher orders. The angle of diffraction depends on the wavelength and the pitch—or periodicity—of the ribbons (Figure 5.13).

The entire display element consists of a two-dimensional array of square pixels, each approximately 20 μm on a side, containing two fixed and two flexible ribbons. The mechanical structure of the ribbon relies on a thin, silicon nitride film under tension to provide the restoring force in the absence of actuation. The reflecting surface is a 50-nm-thick aluminum layer. The underlying electrode is made of tungsten, isolated from the substrate by silicon dioxide.

The optical projection system includes an aperture mounted over the display element. The aperture blocks the reflected light but allows the first diffraction orders to be imaged by the projection lens. The incident

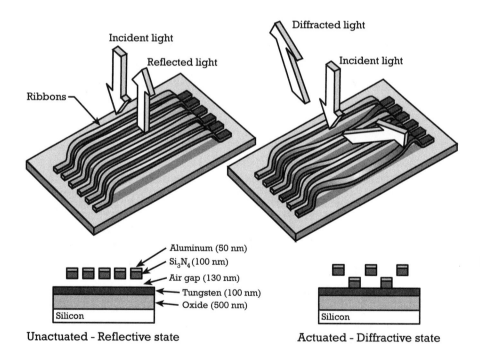

Figure 5.13 Illustration of the operating principle of a single pixel in the grating light valve. Electrostatic pull-down of alternate ribbons changes the optical properties of the surface from reflective to diffractive. Adapted from D. M. Bloom [20].

The New Gearbox: A Peek Into the Future

illumination is normal to the chip, sending the diffracted orders off-axis. Alternatively, the use of off-axis illumination simplifies the imaging optics in a scheme similar to projection with the DMD™, described in the previous chapter.

For full color display, each pixel consists of three sets of ribbons, one for each of the three primary colors: red, green, and blue. The design of the pitch is such that the diffraction order of only a single color from each subpixel is imaged by the projection lens. The pitch of the red subpixel must be larger than that of green, which is in turn larger than that of blue (Figure 5.14).

The GLV™ display supports at least 256 gray shades or 8-bit color depth by rapidly modulating the duration ratio of bright to dark states. This, in turn, varies the light intensity available for viewing—similar to the scheme used in the DLP™ by Texas Instruments (see Chapter 4). Early display prototypes demonstrated a contrast ratio between the bright and dark states in excess of 200. The fill ratio—the percentage area active in reflecting light—is approximately 70%, with a potential for further improvement by reducing the spacing between ribbons. The pitch, and not the spacing, determines the diffraction angle.

A key advantage of the GLV™ over other display technologies is its fast speed. The small size and weight of the ribbon, combined with the

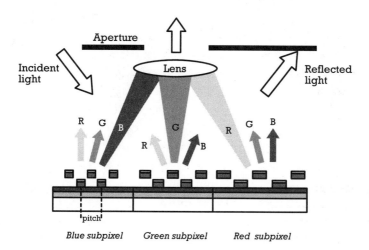

Figure 5.14 Implementation of color in a GLV™ pixel. The pitch of each color subpixel is tailored to steer the corresponding light to the projection lens. The size of the pixel is exaggerated for illustration purposes.

short stroke, provide a switching speed of about 20 ns, about one thousand times faster than the DMD™. At these speeds, the address and support electronics become simple. There is no longer a need for fast buffers, such as those required for conventional active matrix liquid-crystal displays, to compensate for the mismatch in speeds between the electronics and the display elements. Moreover, there is little power required to actuate the very small ribbons.

The very fast switching has also allowed Silicon Light Machines to explore a new scheme whereby the projected image of a single row of pixels is rapidly scanned through the optics to build a two-dimensional picture. Projection at video rate for a high-resolution display requiring 1000 horizontal lines implies a data-scan rate of 60,000 lines per second. Incorporating 256 shades of gray increases the bit-refresh rate to 15.4 MHz, which corresponds to a pixel switching every 65 ns, well within the capability of the GLV™. This new scheme allows simplifying the GLV™ to a single row of pixels instead of a two-dimensional array, and hence reduces associated manufacturing costs.

The fabrication involves the surface micromachining of the ribbons and their release by etching a sacrificial layer. The process begins with the deposition over a silicon wafer of an insulating 500-nm-thick silicon dioxide layer followed by the sputter-deposition or CVD of tungsten. The tungsten is patterned using standard lithography, and etched in SF_6-based plasma to define the electrodes for electrostatic actuation. The sacrificial layer is then deposited. The details of this layer are not publicly available, but there exist many possibilities, including organic polymers. This layer is very thin, measuring approximately 130 nm, one-quarter the wavelength of green light. Silicon nitride and aluminum are deposited next, followed by patterning in the shape of narrow ribbons. The release step is last. Oxygen plasma is useful for the removal of organic sacrificial layers, such as photoresist. It is also possible to consider using sputtered amorphous silicon as a sacrificial layer. Its selective removal, however, may require an exotic etch-step involving xenon difluoride (XeF_2). This etchant sublimes at room temperature from its solid form, and reacts spontaneously with silicon to form SiF_4. Its advantage over SF_6 or CF_4 is that it does not require a plasma and it does not etch silicon nitride, silicon oxide, or aluminum. But xenon difluoride is a hazardous chemical, reacting with water moisture to form hydrofluoric acid. It is not used in the integrated circuit industry.

Optical switches

Few doubt the potential applications of miniature optical switches, whether micromachined or otherwise. They are to optical communications and light transmission what the transistor is to electronic signaling. At least two general application areas have emerged: one in fiber optic communications and another in data storage. Arrays of optical switches allow the rapid reconfiguration of optical networks in data communications by altering the light path in a system of intersecting fibers—much like railroad points move train tracks to reconfigure a rail network. They are also useful components in the addition and deletion of extra channels in optical add/drop multiplexers (OADM) [21] for wavelength division multiplexing (WDM)—equivalent to traffic lights on highway entrance and exit ramps now common throughout California. Switch applications in optical communications systems are truly emerging, with their details often shrouded in secrecy, as innovative start-up companies, with generous funding from private investors and venture capitalists, race against the giants of the telecommunications industry to develop the next optical-switch array.

In data-storage applications, miniature optical switches can steer light pulses to appropriate locations on the platter of a magneto-optical disk. At least one company, Quinta Corporation, San Jose, California, a subsidiary of Seagate Technology, Inc., is exploring the use of micromachined mirrors and optical switches to achieve high-storage densities in excess of 20 Gbit/in^2. The details of this new scheme, dubbed "Optically Assisted Winchester Technology," are proprietary to the company.

Despite the lack of publicly available technical information on micromachined optical switches and their applications, one may gain significant insight by examining the activities at research laboratories and universities, particularly the University of California, Los Angeles, California [22]; and the University of Neuchâtel in Switzerland [23].

A key characteristic of optical switch arrays is their order—in other words, how many input and output fibers can be independently coupled to each other. If a switch can route the light from a single input fiber to any of N output fibers, then it is labeled $1 \times N$. Generally, $M \times N$ switches are two-dimensional arrays, with M input and N output fibers. Their electronic equivalent is an analog multiplexer that selects any one of M electrical inputs and routes its signal to any one of N output lines. Commercially available optical switch arrays from companies such as

E-Tek Dynamics, San Jose, California; and DiCon Fiberoptics, Inc., Berkeley, California; are typically limited to 1 × 2 or 2 × 2. MEMS-based optical switches promise to deliver 64 × 64 arrays and larger. The objective then becomes to demonstrate individual binary-state optical switches that can be readily arrayed to form complex, but miniature, optical multiplexing systems.

The demonstration from the University of Neuchâtel illustrates the basic structure of one possible implementation of a 2 × 2 optical switch using a process somewhat similar to the silicon fusion bonding and deep-reactive-ion etching (SFB-DRIE) process introduced in Chapter 3. It consists of an electrostatic comb actuator controlling the position of a vertical mirror plate at the intersection of two perpendicular grooves; within each lay two optical fibers, one on each groove end. If the mirror plate is retracted, light passes through unobstructed—this is the bar-state. Positioning the plate in the middle of the intersection reflects the light by 90° thereby altering the path of data communication—this is the cross-state (Figure 5.15).

The grooves must normally accommodate optical fibers, typically 150 to 250 μm in diameter. The depth of the grooves must be such that the center of the fiber aligns with the center of the micromirror. The mirror must collect all the light from an individual fiber, and thus should cover the entire fiber core—a central area about 10 μm in diameter that carries light. In the demonstration from the University of Neuchâtel, the mirror height is identical to the depth of the groove, approximately 75 μm. Insertion-loss, a measure of the light-coupling efficiency between input and output fibers, depends on the alignment accuracy of the fibers, with respect to each other and to the mirror. Insertion-loss also relies on the mirror reflecting all the light impinging on its surface. This essentially requires the use of highly reflective coatings, in particular aluminum for wavelengths in the visible, and gold in the infrared. Furthermore, the surface of the mirror must be optically flat in order to eliminate any deleterious light-scattering effects. The operating wavelength in optical fiber communications systems is in the neighborhood of 1.3 μm.

The device from the University of Neuchâtel was fabricated on silicon-on-insulator (SOI) wafers with a 75-μm-thick top silicon layer. Lithography in standard resist was followed by deep-reactive-ion-etching down to the buried oxide. An etch-step in hydrofluoric acid removes the 2 μm buried silicon dioxide layer to release the comb actuator as well as

The New Gearbox: A Peek Into the Future

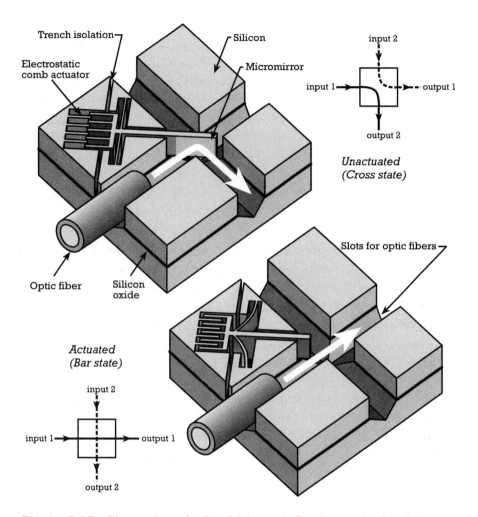

Figure 5.15 Illustration of a 2 × 2 binary reflective optical switch fabricated using silicon-fusion bonding and deep-reactive-ion etching. An electrostatic comb actuator controls the position of a micromirror. In the cross-state, light from an input fiber is deflected by 90°. In the bar-state, the light from that fiber travels unobstructed through the switch. Side schematics illustrate the signal path for each state.

the mirror plate. Finally, 50-nm-thick aluminum is deposited on the silicon surfaces to increase reflectivity.

The optical switch demonstrated an insertion loss of less than 1.6 dB in the bar-state, and less than 3.4 dB in the cross-state. The surface roughness of the mirror plate was 36 nm rms, as measured using atomic force microscopy. The measured reflectivity of the mirror was 76%, lower than the theoretical value of 95%. Scattering from surface roughness, as well as misalignment, are the major contributors to loss of light. The switching time of the device was 200 μs. The silicon area usage is small, measuring about 1 to 2 mm^2. Significant improvements remain to be realized before such a switch can meet the stringent requirements of the telecommunications industry. Nevertheless, it markedly illustrates the potential of MEMS technology in the fabrication of optical switch arrays.

Micropumps

Micropumps are conspicuously missing from the limelight in the United States. In contrast, they receive much attention in Europe and Japan, where the bulk of the developmental activities appear to be. A primary application for micropumps is likely to be in the automated handling of fluids for chemical analysis and drug-delivery systems.

Stand-alone micropump units face significant competition from traditional solenoid or stepper motor-actuated pumps. For instance, The Lee Company, Westbrook, Connecticut, manufactures a family of pumps measuring approximately 51 mm × 12.7 mm × 19 mm (2 in. × 0.5 in. × 0.75 in.) and weighing, fully packaged, a mere 50 g (1.8 oz). They can dispense up to 6 mL/min with a power consumption of 2 W from a 12-V DC supply. But micromachined pumps can have a significant advantage if they can be readily integrated along with other fluid-handling components, such as valves, into one completely automated miniature system. The following demonstration from the Fraunhofer Institute for Solid State Technology, Munich, Germany [24], illustrates one successful effort at making a bidirectional micropump with reasonable flow rates (Figure 5.16).

The basic structure of the micropump is rather simple, consisting of a stack of four wafers. The bottom two wafers define two check valves at the inlet and outlet. The top two wafers form the electrostatic actuation unit. The application of a voltage between the top two wafers actuates the pump diaphragm, thus expanding the volume of the pump inner chamber. This draws liquid through the inlet-check-valve to fill the additional chamber volume. When the applied AC voltage goes through its null point, the diaphragm relaxes and pushes the drawn liquid out through

The New Gearbox: A Peek Into the Future

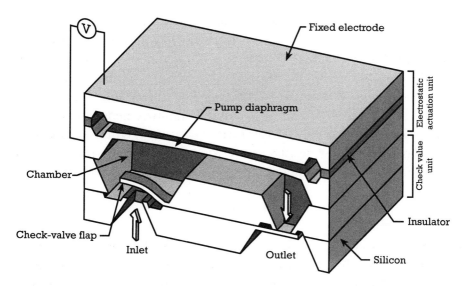

Figure 5.16 Illustration of a cut-out of a silicon micropump from the Fraunhofer Institute for Solid State Technology, Munich, Germany [24]. The overall device measures $7 \times 7 \times 2$ mm^3. The electrostatic actuation of a thin diaphragm modulates the volume inside a chamber. An increase in volume draws liquid through the inlet-check-valve. Relaxation of the diaphragm expels the liquid through the outlet-check-valve.

the outlet-check-valve. Each of the check valves comprises a flap that can move only in a single direction: The flap of the inlet-check-valve moves only as liquid enters to fill the pump inner chamber; the opposite is true for the outlet-check-valve.

The novelty of the design is in its ability to pump fluid in either a forward or reverse direction—hence its bidirectionality. At first glance, it appears that such a scenario is impossible because of the geometry of the two check valves. This is true as long as the pump diaphragm displaces liquid at a frequency lower than the natural frequencies of the two check valve flaps. But at higher actuation frequencies—above the natural frequencies of the flap—the response of the two flaps lags the actuation drive. In other words, when the pump diaphragm actuates to draw liquid into the chamber, the inlet-valve flap cannot respond instantaneously to this action and remains closed for a moment longer. The outlet-check-valve is still open from the previous cycle and does not respond quickly to closing. In this instance, the outlet-check-valve is open and the inlet-check-valve is closed, which draws liquid into the chamber through the outlet rather than the inlet. Hence, the pump reverses its direction. Clearly, for this to happen, the response of the check valves must lag the actuation by at least one quarter of a cycle—the phase difference between the check valves and the actuation must exceed 90°. This occurs at frequencies above the natural frequency of the flap. If the drive frequency is further increased, then the displacement of the flaps becomes sufficiently small that the check valves do not respond to actuation.

The pump rate initially rises with frequency and reaches a peak flow rate of 800 μL/min at 1 kHz. As the frequency continues to increase, the time lag between the actuation and the check valve becomes noticeable. At exactly the natural frequency of the flaps (1.6 kHz), the pump rate precipitously drops to zero. At this frequency, the phase difference is precisely 180°, meaning that both check valves are simultaneously open—hence no flow. The pump then reverses direction with further increase in frequency, reaching a peak backwards flow rate of –200 μL/min at 2.5 kHz. At about 10 kHz, the actuation is much faster than the response of the check valves, and the flow rate is zero. For this particular device, the separation between the diaphragm and the fixed electrode is 5 μm, the peak actuation voltage is 200 V, and the power dissipation is less than 1 mW. The peak hydrostatic back pressure developed by the pump at zero flow is 31 kPa (4.5 psi) in the forward direction, and 7 kPa (1 psi) in the reverse direction.

The fabrication is rather complex involving etching many cavities separately in each wafer, and then bonding the individual substrates together to form the stack. Etching using any of the alkali hydroxides is sufficient to define the cavities. The final bonding can be done either by gluing the different parts or using silicon-fusion bonding (Figure 5.17).

Thermomechanical data storage

Imagine that each storage bit on the surface of the platter of a hard disk is a mere 100 nm on a side, and that the read/write head is a sharp tip able to distinguish this small bit—then a disk area of one square-inch would be able to hold 65 Gbits worth of data, at least one-order-of-magnitude larger than current state-of-the-art densities, and large enough to store ten copies of the Encyclopedia Britannica on a single hard drive. This is

The New Gearbox: A Peek Into the Future

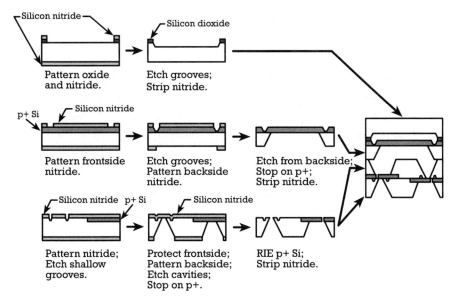

Figure 5.17 Fabrication process for an electrostatically actuated micropump.

precisely what a team of scientists from IBM Almaden Research Center, San Jose, California, and Stanford University is exploring [25].

The idea of using micromachined cantilevers for data storage has continued to be studied ever since atomic-force microscopy (AFM) was invented in the 1980s. In this write-once, read-many-times scheme, a sharp tip on the cantilever head locally alters a physical property on the surface of a spinning disk to encode—or write—data. An obvious property change is to mechanically indent the surface. The data reading involves measuring the presence or absence of pits using high-resolution AFM.

The demonstration from IBM and Stanford University provides a writing cantilever with a heater at its tip to locally melt the surface of a spinning polycarbonate disk. It also eliminates the overhead associated with the laser read-out in a conventional AFM set-up by replacing the sensing with a piezoresistive cantilever capable of measuring slight depressions in the disk surface. The writing and reading tips remain in light contact with the surface of the rotating disk by means of a slight loading force applied to the base of the cantilevers. The loading force is sufficiently small to avoid wear (Figure 5.18).

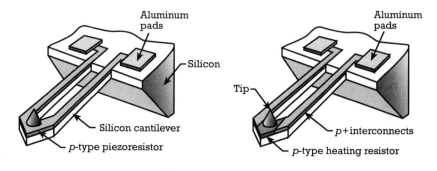

Figure 5.18 Write and read cantilevers with sharp tips for high-density thermomechanical data storage applications. In operation, the sharp tips are in light contact with the surface of a spinning polycarbonate disk. A heater integrated at the tip of the write head locally melts and pits the surface of the disk. The piezoresistive read cantilever measures minute surface indentations that represent permanently recorded digital data [25].

Though the two cantilevers differ in functionality, their structural appearances are very similar. They both consist of two thin arms joined at one end. A sharp tip at the joint provides mechanical contact between the cantilevers and the surface of the disk. In the writing cantilever, a low-doped, p-type layer ($\sim 10^{17}$ cm^{-3}) acts as a resistive heater to raise the tip temperature to 120° C and locally melt the polycarbonate disk. Heavy p-type doping of the arms provides low electrical resistance contacts to the heater. For the reading cantilever, a uniform p-type layer on one surface of the cantilever forms a piezoresistor to sense tip deflections out-of-the-plane. The reading cantilever needs to be very compliant in order to translate depressions only 10 to 20 nm deep into a measurable stress in the piezoresistive element. Moreover, its resonant frequency must be sufficiently high so that the data-reading rates are meaningful. These conditions imply that the cantilever must be very thin, and its mass must be nearly negligible. In the design from IBM and Stanford University, the cantilevers are 1-μm-thick, 75-μm-long, and 10-μm-wide; providing a stiffness of 1.6 N/m, a mass of 5.2 × 10^{-13} kg, and a corresponding natural frequency of 280 kHz.

Writing cantilevers successfully demonstrated a writing rate of 100 kbit/s on a rotating polycarbonate disk using 16-V pulses for a duration of

20 μs. The reading bandwidth was even higher (300 kbit/s). The beam compliance and the doping level of the piezoresistive sense element determine the reading sensitivity given by the relative change in the cantilever resistance ($\Delta R/R$). For this particular device, the relative change in resistance was 8×10^{-6} for every nanometer of tip deflection. Such a small value requires sensitive electronics to detect, amplify, and filter the signal. The detection resolution over the entire bandwidth was 1 nm, limited by $1/f$ and thermal noise. While this may be unacceptable for microscopy, it is adequate to measure the 20-nm indentations made by the writing cantilever(Figure 5.19).

The fabrication of the tip involves the plasma-etching of a 5-μm-thick silicon layer on a silicon-on-insulator (SOI) substrate. The etching in SF_6 is isotropic and removes silicon, both laterally and vertically around a small disk, leaving behind a sharp tip. Oxidation further sharpens the tip and reduces the thickness of the silicon to 1 μm. Subsequent boron

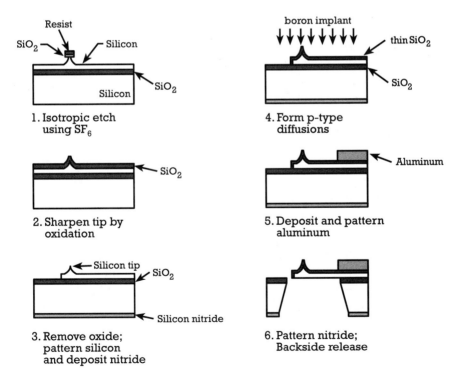

Figure 5.19 Fabrication sequence of reading and writing cantilevers for thermomechanical data storage [25].

implants form the heater and the piezoresistive sense element. Deposition and patterning of aluminum provides electrical contact to the heater and piezoresistor. A final back side etch in tetramethyl ammonium hydroxide (TMAH), followed by an oxide-etch in hydrofluoric acid, releases the cantilevers (Figure 5.20).

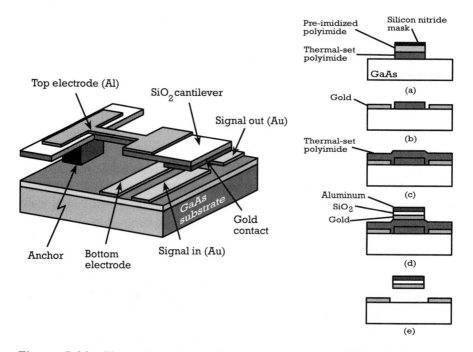

Figure 5.20 Illustration of a surface-micromachined RF-switch and its fabrication sequence over a semi-insulating GaAs substrate. Electrostatic actuation bends the cantilever to close the signal path between two gold lines. (a) Thermal-setting and pre-imidized polyimide layers are deposited. Silicon nitride is then deposited, patterned, and used as a mask to etch the polyimide layers. (b) Gold is evaporated, and the pre-imidized polyimide layer is dissolved. (c) Another polyimide layer is deposited. (d) Gold contact is deposited and patterned. Silicon dioxide is deposited and patterned in the form of a cantilever. Top aluminum electrode is deposited and patterned. (e) Isotropic plasma-etch in oxygen removes all the polyimide layers and releases the cantilever beam. Adapted from Yao and Chang [27].

RF switch over gallium arsenide

The focus of the present and previous chapters was entirely on silicon-based MEMS. Therefore, a conclusion departing from the general trend would normally be unorthodox. But in the context of introducing MEMS to the general reader, a departure from silicon is appropriate to emphasize the applicability of this technology across a broad range of materials.

Operation in the radio frequency (RF) range above 1 GHz is intimately linked to electronic devices made of gallium arsenide (GaAs) or other group III-V compound semiconductors. Unfortunately, electronic switches, such as GaAs MESFETs, do not provide a high degree of isolation in the open state, and they suffer from significant insertion losses in the closed state. A MEMS-based RF-switch could be an appropriate substitute [26].

The following device from the Rockwell Science Center, Thousand Oaks, California, demonstrates an RF-switch with operation from DC up to 4 GHz [27]. It consists of a suspended silicon dioxide cantilever over a semi-insulating GaAs substrate (Figure 5.20). This type of substrate is advantageous over silicon or other semiconductors because it reduces the capacitive coupling between adjacent signal lines at high frequencies, and thus improves the electrical isolation in the open state. Moreover, the use of GaAs substrates permits the integration of the RF switch with high-speed electronic devices, such as MESFETs.

In the closed state, an applied voltage between the cantilever and an electrode on the substrate pulls a gold contact down, shorting the input and output signal lines. A cantilever 100- to 200-μm-long, 10-μm-wide, and 2-μm-thick, separated from the substrate by a 3 μm gap, closes in approximately 30 μs, with an actuation voltage of 28 V. The measured insertion loss and electrical signal isolation at 4 GHz were 0.1 dB and –50 dB, respectively.

The fabrication entails the use of organic polymers as sacrificial layers that are later removed to mechanically release the cantilever (Figure 5.20). All process steps require temperatures of 250° C or less in order to integrate the switch with millimeter-wave-integrated circuits (MMIC). First, a thermal setting polyimide is spin-deposited and cured. A second pre-imidized polyimide layer is also spin-deposited and cured. Silicon nitride is then deposited and patterned using standard lithography and plasma-etching in CHF_3. Etching using oxygen plasma transfers the pattern into the polyimide layers. Gold is then evaporated. Dissolution in methylene chloride removes the pre-imidized polyimide and all layers on top of it. This lift-off process leaves behind the first polyimide layer and the gold in contact with the substrate. Another polyimide layer is spun over the substrate and cured. A layer of gold, 1-μm-thick, is evaporated, and patterned to define the gold contact. This is followed by the deposition using PECVD of 2-μm-thick silicon dioxide and its patterning in the form of the cantilever. A final evaporation step deposits 0.25-μm-thick aluminum that is subsequently patterned in the shape of an electrode. Finally, an oxygen isotropic plasma-etch step removes all the polyimide layers and releases the cantilever beam.

Summary

The future promises to bring innovative and novel MEMS solutions to a wide variety of applications, ranging from biochemical analysis to wireless and optical systems. We reviewed in this chapter a series of devices that illustrate this diversity. Most of these devices and systems remain in the research and development phase, but show significant potential for becoming commercial products.

References

[1] Wu, M. C., "Micromachining for Optical and Optoelectronic Systems," *Proceedings of the IEEE*, Vol. 85, No. 11, Nov. 1997, pp. 1833–1856.

[2] Muller, R. S., and K. Y. Lau, "Surface-Micromachined Microoptical Elements and Systems," in *Integrated Sensors, Microactuators, & Microsystems (MEMS)*, pp. 1705–1720, K. D. Wise (ed.), Proceedings of the IEEE, Vol. 86, No. 8, Aug. 1998.

[3] Pister, K. S. J., et al., "Microfabricated Hinges," *Sensors and Actuators*, Vol. A33, No. 3, June 1992, pp. 246–256.

[4] Mastrangelo, C. H., M. A. Burn, and D. T. Burke, "Microfabricated Devices for Genetic Diagnostics," in *Integrated Sensors, Microactuators, & Microsystems (MEMS)*, pp. 1769–1787, K. D. Wise (ed.), Proceedings of the IEEE, Vol. 86, No. 8, Aug. 1998.

[5] Stryer, L., *Biochemistry*, New York, NY: W. H. Freeman and Co., 1988, pp. 71–90, 120–123.

[6] Darnell, J., L. Harvey, and D. Baltimore, *Molecular Cell Biology*, 2nd ed., New York, NY: Scientific American Books, 1990, p. 219.

[7] Northrup, M. A., et al., "DNA Amplification with a Microfabricated Reaction Chamber," *Proc. 7th Int. Conf. on Solid-State Sensors and Actuators*, Yokohama, Japan, June 7–10, 1993, pp. 924–926.

[8] Kuhr, W. G., and C. A. Monnig, "Capillary Electrophoresis," *Analytical Chemistry*, Vol. 64, 1992, pp. 389R–407R.

[9] Manz, A., et al., "Planar Chips Technology for Miniaturization and Integration of Separation Techniques into Monitoring Systems. Capillary Electrophoresis on a Chip," *Journal of Chromatography*, Vol. 593, 1992, pp. 253–258.

[10] Woolley, A. T., and R. A. Mathies, "Ultra-High Speed DNA Sequencing Using Capillary Electrophoresis Chips," *Analytical Chemistry*, Vol. 67, 1995, pp. 3676–3680.

[11] Kovacs, G. T. A., "Introduction to the Theory, Design, and Modeling of Thin-Film Microelectrodes for Neural Interfaces." In *Enabling Technologies for Cultured Neural Networks*, pp. 121–166, D. A. Stenger and T. M. McKenna (eds.), San Diego, CA: Academic Press, 1994.

[12] U.S. Patents #5,605,662 (Feb. 25, 1997) and #5,632,957 (May 27, 1997).

[13] Borkholder, D. A., B. D. DeBusschere, and G. T. A. Kovacs, "An Approach to the Classification of Unknown Biological Agents with Cell Based Sensors," *Tech. Digest Solid-State Sensor and Actuator Workshop*, Hilton Head Island, SC, June 8–11, 1998, pp. 178–182.

[14] Wang, J., *Stripping Analysis: Principles, Instrumentation, and Applications*, Deerfield Beach, FL: VCH, 1985.

[15] Kovacs, G. T. A., C. W. Storment, and S. P. Kounaves, "Microfabricated Heavy Metal Ion Sensor," *Sensors and Actuators*, Vol. B23, 1995, pp. 41–47.

[16] Tsai, S., et al., "Novel Analytical Technique for On-Line Monitoring of Trace Heavy Metals in Corrosive Chemicals." In *Characterization and Metrology for ULSI Technology*, pp. 907–912, D. G. Seiler, A. C. Diebold, W. M. Bullis, T. J. Shaffner, R. McDonald, and E. J. Walters (eds.), New York, NY: The American Institute of Physics, 1998.

[17] Nguyen, C. T.-C., "Frequency-Selective MEMS for Miniaturized Communications Devices," *Proc. 1998 IEEE Aerospace Conference*, Vol. 1, Snowmass, Colorado, Mar. 21–28, 1998, pp. 445–460.

[18] Kittel, C., *Introduction to Solid State Physics*, 6th edition, New York, NY: Wiley, 1986, pp. 82–91.

[19] Wang, K., and C. T.-C. Nguyen,, "High-Order Mechanical Electronic Filters," *Proc. IEEE Micro Electro Mechanical Systems*, Nagoya, Japan, Jan. 26–30, 1997, pp. 25–30.

[20] Bloom, D. M., "The Grating Light Valve: Revolutionizing Display Technology," *Proc. SPIE, Projection Displays III*, Vol. 3013, San Jose, CA, Feb. 10–12, 1997, pp. 165–171.

[21] Agrawal, G. P., *Fiber-Optic Communication Systems*, 2nd ed., New York, NY: Wiley, 1997.

[22] Lee, S.-S., et al., "2 × 2 MEMS Fiber Optic Switches with Silicon Sub-Mount for Low-Cost Packaging," *Tech. Digest Solid-State Sensor and Actuator Workshop*, Hilton Head Island, SC, June 8–11, 1998, pp. 281–284.

[23] Marxer, C., et al., "Vertical Mirrors Fabricated by Deep Reactive Ion Etching for Fiber-Optic Switching Applications," *Journal of Microelectromechanical Systems*, Vol. 6, No. 3, Sept. 1997, pp. 277–185.

[24] Zengerle, R., et al., "A Bidirectional Silicon Micropump," *Proc. IEEE Micro Electro Mechanical Systems*, Amsterdam, the Netherlands, Jan. 29–Feb. 2, 1995, pp. 19–24.

[25] Chui, B. W., et al., "Low-stiffness Silicon Cantilevers for Thermal Writing and Piezoresistive Readback with the Atomic Force Microscope," *Applied Physics Letters*, Vol. 69, No. 18, 28 Oct. 1996, pp. 2767–2769.

[26] De Los Santos, H. J., *Introduction to Microelectromechanical (MEM) Microwave Systems*, Norwood, MA: Artech House, 1999.

[27] Yao, J. J., and M. F. Chang, , "A Surface Micromachined Miniature Switch for Telecommunications Applications with Signal Frequencies from DC up to 4 GHz," *Proc. 8th Int. Conf. on Solid-State Sensors and Actuators*, Stockholm, Sweden, June 25–29, 1995, pp. 384–387.

Selected bibilography

Kovacs, G. T. A., *Micromachined Transducers Sourcebook*, New York, NY: McGraw-Hill, 1998.

Horton, R. M., and R. C. Tait, *Genetic Engineering with PCR*, Norfolk, UK: Horizon Press, 1998.

"The E-Nose. Silicon Scents a Need," *IEEE Spectrum*, Special Report on Electronic Noses, Sept. 1998, pp. 22–38.

Wise, K. D., Editor, "Special Issue on Integrated Sensors, Microactuators, and Microsystems (MEMS)," *Proceeding of the IEEE*, Vol. 86, No.8, Aug. 1998.

CHAPTER 6

Contents

Key design and packaging considerations

Die-attach processes

Wiring and interconnects

Types of packaging solutions

Summary

The Box: Packaging for MEMS

Things derive their being and nature by mutual dependence and are nothing in themselves.

Nagarjuna, Indian Buddhist philosopher, ca AD 200. Quoted in the "Central Philosophy of Buddhism," by T. R. V. Murti.

Packaging is the process, industry, and methods of "packing" microelectromechanical components and systems inside a protective housing. Combining engineering and manufacturing technologies, it converts a micromachined structure or system into a *useful* assembly that can *safely* and *reliably* interact with its surroundings. The definition is broad because each application is unique in its packaging requirements. In the integrated circuit industry, electronic packaging must provide reliable, dense interconnections to the multitude of high-frequency electrical signals. In contrast, MEMS packaging must account for a far more complex and diverse set of parameters. It must first protect the micromachined parts in broad-ranging environments; it must also provide interconnects to electrical signals, and, in some cases, access to and interaction with the external environment. For example, the packaging of a pressure sensor must ensure that the sensing device is in intimate contact with the pressurized medium, yet protected from exposure to any harmful substances in this medium. Moreover, packaging of valves must provide both electrical and fluid interconnects. As a consequence of these diverse requirements, standards for MEMS packaging are lacking and designs often remain proprietary to companies. Invariably, the difficulty and failure in adopting standards implies that packaging will remain engineering-resource-intensive, and thus will continue to carry rather high-fixed costs.

Packaging is a necessary evil. Its relatively large dimensions tend to dilute the small size advantage of MEMS. It is also expensive: the cost of packaging tends to be significantly larger than the cost of the actual micromachined components. It is not unusual that the packaging content is responsible for 75% to 95% of the overall cost of a microelectromechanical component or system. These factors, prevalent in the early days of electronic integrated circuits, contributed toward large-scale integration in that industry, in order to minimize the impact of packaging on overall cost and size. High-density packaging methods, such as surface-mount technologies (SMT), are today at the core of advancements in electronic packaging. In contrast, the evolution of MEMS packaging is slow and centers largely on borrowing from the integrated circuit industry in an effort to benefit from the existing vast body of knowledge. Whether sophisticated packaging technologies will penetrate MEMS remains to be seen. If they do, they will certainly have to rely on serious market incentives, particularly high-volume applications, and on a minimum level of technology standardization.

The field of packaging is so broad in scope that one can only hope to present here a brief introduction of the basic fundamentals, especially as they relate to the various structures and systems introduced in the previous chapters. Such an accomplishment is made more difficult by the proprietary nature of most package designs (Figure 6.1).

Key design and packaging considerations

Designing packages for micromachined sensors and actuators involves taking into account a number of important factors. Some of these are

The Box: Packaging for MEMS

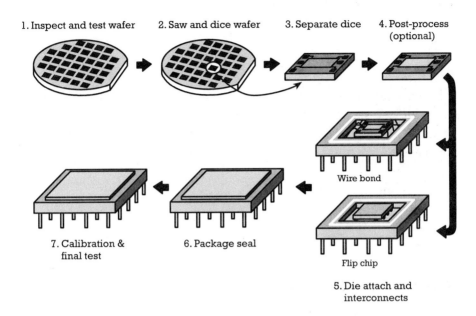

Figure 6.1 Illustration of a simplified process flow for MEMS packaging. Upon completion of wafer-level fabrication, inspection and first tests take place. The wafer is then mounted on a special sticky tape and sawed. The individual dice are separated. Some postprocessing, such as removal of a sacrificial layer, may occur at this point. One or more dice are attached to a ceramic, a metal header, or a premolded plastic lead frame. Electrical interconnects are made by wire bonding, flip-chip, or another method. A ceramic, metal, or plastic cap seals the assembly. Alternatively, the die or dice are attached to a metal lead frame. After the electrical interconnects are made, plastic is molded over the assembly. A final test and calibration conclude the process.

shared with the packaging of electronic integrated circuits, but many are specific to the application. These factors also bear significance to the design of the micromachined components themselves. As a result, the design of the package and of the micromachined structures must commence and evolve together. It would be naïve to believe that they can be separated. The following are critical factors and considerations frequently encountered in MEMS packaging.

Wafer or wafer-stack thickness

Standards in the electronic integrated circuit industry dictate specific thicknesses for silicon wafers, depending on their diameters. For example, a standard 100-mm (4-in.) diameter silicon wafer polished on one side has a nominal thickness of 525 mm. The standard thickness increases to 650 μm for 150-μm-diameter (6-in.) wafers. Wafers polished on both sides are normally thinner. Glass substrates are at least 250 μm (10 mils) thick. Often, a stack of bonded silicon or glass wafers can have a total thickness exceeding 1 mm, posing significant challenges for packaging facilities. In some cases, it becomes outright impossible to accommodate such large thicknesses. Proper communication of the thickness to the parties responsible for packaging is imperative in order to minimize disruptions to the assembly line and avoid unnecessary delays.

Wafer dicing concerns

A key highlight of MEMS technology is the batch fabrication aspect—hundreds and thousands of identical structures or microsystems are fabricated simultaneously on the same wafer. Dicing separates these structures into individual components (dice) that can be packaged later. A diamond or carbide saw blade, approximately 75- to 250-μm-wide, spins at high speed and cuts through the substrate, which is normally mounted and held in position on a blue-colored "sticky tape." Water flows continuously during sawing to cool the blade. Dicing is a harsh process conducted in an unclean environment and subjects the microstructures to strong vibrations and shaking. Retaining the integrity and cleanliness of the microstructures requires protecting the sensitive components from particulates and liquids, as well as ensuring that they can survive all the shaking.

Each MEMS design merits its own distinctive approach on how to minimize the adverse effects of dicing. In surface-micromachined MEMS, such as the accelerometer from Analog Devices, protection can mean, for example, forming shallow dimples in the blue sticky tape and mounting the wafer upside down, so that the sensitive micromechanical structures face towards and are aligned with the dimples. Alternatively, it is possible to perform the final sacrificial etch (see Chapter 3) *after* the dicing is complete. While this "postprocess" approach

The Box: Packaging for MEMS

ensures there are no free mechanical structures during the dicing, it implies that the microstructures must be freed on each individual die, thus sacrificing batch fabrication for mechanical integrity. This naturally increases the final fabrication cost. The fabrication process of the Texas Instruments Digital Mirror Device (DMD™) follows a similar approach. The DMD™ arrays are diced first, then the organic sacrificial layer is consequently etched in an oxygen plasma. Since the rumored selling price for each DMD™ is in the hundreds of dollars, this method may be economically justified. But accelerometers intended for the automotive market command prices of a few dollars at most, with little margin to allocate to the dicing process.

The reader will observe in Chapters 4 and 5 a number of designs incorporating bonded caps or covers made of silicon and occasionally glass, whose sole purpose is to protect the sensitive micromechanical structures. These become, after the completion of the cap, fully embedded inside an all-micromachined housing—a first-level package. For example, the yaw-rate sensor from Robert Bosch GmbH includes a silicon cover that protects the embedded microstructures during dicing, provided the vibrations are not sufficiently large to cause damage. In addition to mechanical protection, an electrically grounded cover also shields against electromagnetic interference (EMI). Naturally, the cap approach is not suitable for sensors, such as pressure or flow sensors, or actuators that require direct and immediate contact with their surrounding environments.

Thermal management

The demands on thermal management can be very diverse and occasionally conflicting, depending on the nature of the application. The main role of thermal management for electronic packaging is to cool the integrated circuit during operation [1]. A modern microprocessor containing millions of transistors and operating at a few hundred megahertz can consume tens of watts. In contrast, the role of thermal management in MEMS includes the cooling of heat-dissipating devices, especially thermal actuators, but also involves understanding and controlling the sources of temperature fluctuations that may adversely affect the performance of a sensor or actuator. As such, thermal management is performed at two levels: the die level and the package level.

Thermal analysis is analogous to understanding electrical networks. This is not surprising because of the dual nature of heat and electricity—voltage, current, and electrical resistance are dual to temperature, heat flux, and thermal resistance, respectively. A network of resistors is an adequate first-order model to understand heat flow and nodal temperatures. The thermal resistance, θ, of an element is equal to the ratio of the temperature difference across the element over the heat flux—this is equivalent to Ohm's law for heat flow. For a simple slab of area A and length l, θ equals $l/(kA)$, where k is the thermal conductivity of the material (Figure 6.2).

The nature of the application severely influences the thermal management at the die level. For example, in typical pressure sensors that dissipate a few tens of milliwatts over an area of several square millimeters, the role of thermal management is to ensure long-term thermal stability of the piezoresistive sense elements by verifying that no thermal gradients arise within the membrane. The situation becomes more complicated if any heat-dissipating elements are positioned on very thin

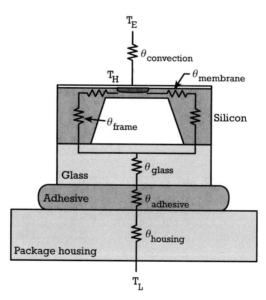

Figure 6.2 Components of thermal resistance for a hypothetical microstructure, including a heat-producing element at temperature T_H, embedded in a suspended membrane. The device is assembled within a housing maintained at a low temperature, T_L. The temperature of the surrounding environment is T_E.

membranes, increasing the effective thermal resistance to the substrate and the corresponding likelihood of temperature fluctuations. Under some circumstances, maintaining an element at a constant temperature above ambient brings performance benefits. One example is the mass-flow sensor from Honeywell (see Chapter 4).

Thermal management at the package level must take into account all the thermal considerations of the die level. In the case of the mass-flow sensor, it is imperative that the packaging does not interfere with the die-level thermal isolation scheme. In the example of the infrared imager, also from Honeywell (see Chapter 4), the package housing needs to hold a permanent vacuum to eliminate convective heat loss from the suspended sensing pixels.

Thermal actuators can dissipate significant power. It takes at least a few watts for a thermal actuator to deliver a force of 100 mN with a displacement of 100 μm. With efficiencies typically below 0.1%, most of the power is dissipated as heat that must be removed through the substrate and package housing. In this case, thermal management shares many similarities with the thermal management of electronic integrated circuits. This is a topic that has been thoroughly studied and published on [1].

Ceramics and metals make excellent candidate materials for the package housing because of their high thermal conductivity. To ensure unimpeded heat flow from the die to the housing, it is necessary to select a die-attach material that does not exhibit a low thermal conductivity. This may exclude silicones and epoxies and instead favor solder-attach methods or silver-filled epoxies. A subsequent section in this chapter explores various die-attach techniques. Naturally, a comprehensive thermal analysis should take into account all mechanisms of heat loss, including loss to fluid in direct contact with the actuator.

Stress isolation

The previous chapters described the usefulness of piezoresistivity and piezoelectricity to micromachined sensors. By definition, such devices rely on converting mechanical stress to electrical energy. It is then imperative that the piezoresistive or piezoelectric elements are not subject to mechanical stress of undesirable origin, and extrinsic to the parameter that needs to be sensed. For example, a piezoresistive pressure sensor gives an incorrect pressure measurement if the package housing subjects the silicon die to stresses. These stresses need only be minute to have a catastrophic effect, because the piezoresistive elements are extremely sensitive to stress. Consequently, sensor manufacturers take extreme precautions in the design and implementation of packaging. The manufacture of silicon pressure sensors, especially those designed to sense low pressures (< 100 kPa), includes the anodic bonding of a thick (1 mm) Pyrex® glass substrate with a coefficient of thermal expansion matched to that of silicon. The glass improves the sensor's mechanical rigidity, and ensures that any stresses between the sensor and the package housing are isolated from the silicon piezoresistors.

Another serious effect of packaging on stress-sensitive sensors is long-term drift resulting from slow creep in the adhesive or epoxy that attaches the silicon die to the package housing. Modeling of such effects is extremely difficult, leaving engineers with the task of constant experimentation to find appropriate solutions. This illustrates the type of "black art" in the packaging of sensors and actuators, and a reason that companies do not disclose their packaging secrets.

Protective coatings and media isolation

Sensors and actuators coming into intimate contact with external media must be protected against adverse environmental effects, especially if the devices are subject to long-term reliability concerns. This is often the case in pressure or flow sensing, where the medium in contact is other than dry air. For example, sensors for automotive applications must be able to withstand salt water and acid rain pollutants (e.g., SO_x and NO_x). In home appliances (white goods), sensors may be exposed to alkaline environments due to added detergents in water. Even humidity can cause severe corrosion of sensor metalization, especially aluminum.

In many instances of mildly aggressive environments, a thin conformal coating layer is sufficient protection. A common material for coating pressure sensors is parylene (poly (*p*-xylylene) polymers) [2,3]. It is normally deposited using a near-room-temperature chemical vapor deposition process. The deposited film is conformal, covering the sensor element and exposed electrical wires. It is resistant to automotive exhaust gases, fuel, salt spray, water, alcohol, and many organic solvents.

However, extended exposure to highly acidic or alkaline solutions ultimately results in the failure of the coating (Table 6.1).

Recent studies suggest that silicon carbide may prove to be an adequate coating material for protecting MEMS in very harsh environments [4]. Silicon carbide deposited in a plasma-enhanced chemical vapor deposition system by the pyrolysis of silane (SiH_4) and methane (CH_4) at 300° C proved to be an effective barrier for protecting a silicon pressure sensor in a hot potassium hydroxide solution, which is a highly corrosive chemical and a known etchant of silicon. However, much development remains to be done to fully characterize the properties of silicon carbide as a coating material.

For extreme environments, such as applications involving heavy industries, aerospace, or oil drilling, special packaging is necessary to provide adequate protection for the silicon microstructures. If the silicon parts do not need to be in direct contact with the surrounding environment, then a metal or ceramic hermetic package may be sufficient. This is adequate for accelerometers, for example, but inappropriate for pressure

Table 6.1
Material Properties of Three Types of Parylene Coatings*

Property	Parylene-N	Parylene-C	Parylene-D
Density (g/cm^{-3})	1.110	1.29	1.418
Tensile modulus (GPa)	2.4	3.2	2.8
Permittivity	2.65	3.15	2.84
Refractive index	1.661	1.639	1.669
Melting point (°C)	410	290	380
Coefficient of expansion (10^{-6}/K)	69	35	< 80
Thermal conductivity (W/m·K)	0.12	0.082	—
Water absorption (%)	0.01	0.06	< 0.1
Gas permeability (amol/Pa·s·m)			
N_2	15.4	2.1	9.0
CO_2	429.0	15.4	26.0
SO_2	3790.0	22.0	9.53

* They are stable at cryogenic temperatures to over 125° C [2].

or flow sensors. Such devices must be isolated from direct exposure to their surrounding media, and yet continue to measure pressure or flow-rate. Clever media isolation schemes for pressure sensors involve immersing the silicon microstructure in a special silicone oil, with the entire assembly contained within a heavy duty steel package. A flexible steel membrane allows the transmission of pressure through the oil to the sensor's membrane. Media-isolated pressure sensors are discussed in further detail later in this chapter.

Media isolation can be more difficult to achieve in certain applications. For instance, there are numerous demonstrations of optical microspectrometers capable of detecting SO_x and NO_x, two components of smog pollution. But incorporating these sensors into the tail pipe of an automobile is very difficult, because the sensor must be isolated from the harsh surrounding environment, yet light must reach the sensor. A transparent glass window is not adequate because of the long-term accumulation of soot and other carbon deposits.

Hermetic packaging

A hermetic package is theoretically defined as one that prevents the diffusion of helium. For small volume packages (< 0.40 cm^{-3}), the leak-rate of helium must be lower than 5×10^{-8} cm^{-3}/s. In practice, it is always understood that a hermetic package prevents the diffusion of moisture and water vapor through its walls. A hermetic package must be made of metal, ceramic, or millimeter-thick glass. Silicon also qualifies as a hermetic material. Plastic and organic-compound packages, on the other hand, may pass the strict helium leak-rate test, but over time they allow moisture into the package interior; hence, they are not considered hermetic. Electrical interconnections through the package must also conform to hermetic sealing. In ceramic packages, metal pins are embedded and brazed within the ceramic laminates. For metal packages, glass firing yields a hermetic glass-metal seal.

A hermetic package significantly increases the long-term reliability of electrical and electronic components. By shielding against moisture and other contaminants, many common failure mechanisms, including corrosion, are simply eliminated. For example, even deionized water can leach out phosphorous from low-temperature oxide (LTO) passivation layers to form phosphoric acid which, in turn, etches and corrodes aluminum wiring and bond pads. The interior of a hermetic package is typically evacuated or filled with an inert gas such as nitrogen, argon, or helium. The Digital Mirror Device™ from Texas Instruments and the infrared imager from Honeywell, both discussed in a previous chapter, utilize vacuum hermetic packages with transparent optical windows. The package for the DMD™ even includes a getter to absorb any residual moisture.

Calibration and compensation

The performance characteristics of precision sensors; especially pressure, flow, acceleration, and yaw-rate sensors; often must be calibrated in order to meet the required specifications. Errors frequently arise due to small deviations in the manufacturing process. For example, the sensitivity of a pressure sensor varies with the square of the membrane thickness. A typical error of ±0.25 μm on a 10-μm-thick membrane produces a ±5% error in sensitivity, that must often be trimmed to less than ±1%. In addition, any temperature dependence of the output signal must be compensated for.

One compensation and calibration scheme utilizes a network of laser-trimmed resistors with near-zero TCR to offset errors in the sensor [5]. The approach employs all passive components, and is an attractive low-cost solution. The resistors can be either thin-film (< 1-μm-thick) or thick-film (~ 25-μm-thick) [6], and are trimmed by laser ablation. Thin-film resistors, frequent in analog integrated circuits such as precision operational amplifiers, are sputtered or evaporated directly on the silicon die and are usually made of nickel-chromium or tantalum-nitride. These materials have a sheet resistance of about 100 to 200 Ω per square, and a TCR of ±0.005% per °C. Nickel-chromium can corrode if not passivated with quartz or silicon monoxide (SiO), but tantalum nitride self-passivates by baking in air for a few minutes. Thick-film resistors, in contrast, are typically fired on thick ceramic substrates and consist of chains of metal-oxide particles embedded in a glass matrix. Ruthenium dioxide (RuO_2) and bismuth ruthenate ($BiRu_2O_7$) are examples of active metal oxides. Blending the metal oxides with the glass in different proportions produces sheet resistances with a range of values from 10 to 10^6 Ω per square. Their TCR is typically in the range of ±0.01% per °C. Trimming using a neodymium-doped yttrium-aluminum-garnet (Nd:YAG) laser at a wavelength of 1.06 μm produces precise geometrical cuts in the thin- or thick-film resistor, hence adjusting its resistance value. The laser is part of a closed-loop system that continuously monitors the value of the resistance and compares it to a desired target value.

Laser ablation is also useful to calibrate critical mechanical dimensions by direct removal of material. For instance, a laser selectively ablates minute amounts of silicon to calibrate the two resonant modes of the Daimler Benz tuning fork yaw-rate sensor (see Chapter 4). Laser ablation can also be a useful process to precisely calibrate the flow of a liquid through a micromachined channel. For some drug delivery applications such as insulin injection, the flow must be calibrated to within ±0.5%. Given the inverse cubic dependence of flow resistance on channel depth, this translates to an etch depth precision of better than ±0.17%, equivalent to 166 nm in a 100-μm-deep channel. This is impossible to achieve using most, if not all, silicon-etching methods. A laser ablation step can control the size of a critical orifice, under closed-loop measurement of the flow, to yield the required precision.

As the integration of circuits and sensors becomes more prevalent, the trend has been to perform, when possible, calibration and compensation electronically. Many modern commercial sensors, including pressure, flow, acceleration, and yaw-rate sensors, now incorporate application-specific integrated circuits (ASICs) to calibrate the sensor's output and compensate any errors. Correction coefficients are stored in on-chip permanent memory such as EEPROM.

The need to calibrate and compensate extends beyond conventional sensors. For example, the infrared imaging array from Honeywell must calibrate each individual pixel in the array and compensate for any manufacturing variations across the die. The circuits perform this function using a shutter: The blank scene, that is the collected image while the shutter is closed, incorporates the variation in sensitivity across the array. While the shutter is open, the electronic circuits subtract the blank scene image from the active image to yield a calibrated and compensated picture.

Die-attach processes

Subsequent to dicing of the substrate, each individual die is mounted inside a package and attached (bonded) onto a platform made of metal or

The Box: Packaging for MEMS

ceramic, though plastic is also possible under limited circumstances. Careful consideration must be given to die attaching because it strongly influences thermal management and stress isolation. Naturally, the bond must not crack over time or suffer from creep—its reliability must be established over very long periods of time. The following section describes die-attach processes common in the packaging of silicon micromachined sensors and actuators. These processes were largely borrowed from the electronics industry.

Generally, die-attach processes employ metal alloys or organic or inorganic adhesives as intermediate bonding layers [7,8]. Metal alloys are comprised of all forms of solder, including eutectic and noneutectic (Table 6.2). Organic adhesives consist of epoxies, silicones, and polyimides. Solders, silicones, and epoxies are vastly common in MEMS packaging. Inorganic adhesives are glass matrices embedded with silver and resin, and are mostly used in the brazing of pressed ceramic packages (e.g., CERDIP- and CERQUAD-type) in the integrated circuits industry. Their utility for die-attach is limited because of the high-temperature (> 400° C) glass seal and cure operation.

The choice of a solder alloy depends on its having a suitable melting temperature, as well as appropriate mechanical properties. A solder firmly attaches the die to the package and normally provides little or no stress isolation when compared with organic adhesives. However, the bond is very robust and can sustain very large, normal pull-forces on the order of 5,000 N/cm^2. The large mismatch in the coefficients of thermal expansion with silicon or glass results in undesirable stresses that can cause cracks in the bond.

Most common solders are binary or ternary alloys of lead (Pb), tin (Sn), indium (In), antimony (Sb), bismuth (Bi), or silver (Ag) (Figure 6.3). Solders can be either hard or soft. Hard solders (or brazes) melt at temperatures near or above 500° C, and are used for lead and pin attachment in ceramic packages. In contrast, soft solders melt at lower temperatures and, depending on their composition, are classified as eutectic or noneutectic. Eutectic alloys go directly from liquid to solid phase, without an intermediate paste-like state mixing liquid and solid—effectively, eutectic alloys have identical solidus and liquidus temperatures. They have the lowest melting points of alloys sharing the same constituents, and tend to be more rigid, with excellent shear strength.

Table 6.2
Properties of Some Eutectic and Noneutectic Solders [7]

	Alloy	Liquidus (° C)	Solidus (° C)	Ultimate Tensile Strength (MPa)	Uniform Elongation (%)	Creep Resistance
Noneutectic	60%In 40%Pb	185	174	29.58	10.7	Moderate
	60%In 40%Sn	122	113	7.59	5.5	Poor—soft alloy
	80%Sn 20%Pb	199	183	43.24	0.82	Moderate
	25%Sn 75%Pb	266	183	23.10	8.4	Poor
	5%Sn 95%Pb	312	308	23.24	26	Moderate to high
	95%Sn 5%Sb	240	235	56.20	1.06	High
Eutectic	96.5%Sn 3.5%Ag	221	221	57.65	0.69	High
	42%Sn 58%Bi	138	138	66.96	1.3	Moderate—brittle alloy
	63%Sn 37%Pb	183	183	35.38	1.38	Moderate
	1%Sn 97.5%Pb 1.5%Ag	309	309	38.48	1.15	Moderate
	88%Au 12%Ge	356	356	—	—	Moderate
	96.4%Au 3.6%Si	370	370	—	—	Moderate

Silicon and glass cannot be directly soldered to, and thus must be coated with a thin metal film to wet the surface. Platinum, palladium, and gold are good choices, though gold is not as desirable with tin-based solders because of leaching. Leaching is the phenomenon by which metal is absorbed into the solder to an excessive degree, causing intermetallic compounds detrimental to long-term reliability—gold or silver will dissolve into a tin-lead solder within a few seconds. Typically, a thin (< 50 nm) layer of titanium is first deposited on the silicon to improve adhesion, followed by the deposition of a palladium or platinum layer, a

The Box: Packaging for MEMS

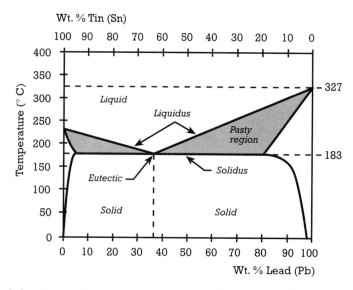

Figure 6.3 Phase diagram of lead-tin solder alloys. The eutectic point corresponds to a lead composition of 37% by weight [7].

few hundred nanometers thick. A subsequent flash-deposition of very thin gold improves surface wetting. Immersing the part in flux (an organic acid) removes metal oxides and furnishes clean surfaces. In a manufacturing environment, the solder paste is either dispensed through a nozzle or screen-printed on the package substrate, and the die is positioned over the solder. Heating in an oven or by direct infrared radiation melts the solder, dissolving, in the process, a small portion of the exposed thin metal surfaces. When the solder cools it forms a joint, bonding the die to the package. Melting in nitrogen or in forming gas prevents oxidation of the solder.

Organic adhesives are attractive alternatives to solder because they are inexpensive, easy to automate, and they cure at lower temperatures. The most widely used are epoxies and silicones, including room-temperature vulcanizing (RTV) rubbers. Epoxies are thermosetting (i.e., cross-linking when heated) plastics with cure temperatures varying between 50° and 175° C. Filled with silver or gold, they become thermally and electrically conductive, but not as conductive as solder. Electrically nonconductive epoxies may incorporate particles of aluminum oxides, beryllium oxides, or magnesium oxides for improved thermal conductivity. RTV silicones come in a variety of specifications for a wide range of applications, from construction to electronics. For example, the Dow Corning® 732 is a multipurpose silicone that adheres well to glass, silicon, and metal, with a temperature rating of −65° C to 232° C [9]. Most RTV silicones are one-part compounds curing at room temperature in air. Unlike epoxies, they are soft and are excellent choices for stress relief between the package and the die. The operating temperature for most organic adhesives is limited to less than 200° C, otherwise they suffer from structural breakdown and outgassing.

Epoxies and RTV silicones are suitable for automated manufacturing. As viscous pastes, they are dispensed by means of nozzles at high rates, or screen-printed. The placement of the die over the adhesive may also be automated by using "pick-and-place" robotic stations employing pattern recognition algorithms for accurate positioning of the die.

Wiring and interconnects

With the advent of microfluidic components and systems, the concept of interconnects is now more global, simultaneously incorporating electrical and fluid connectivity. Electrical connectivity addresses the task of providing electrical wiring between the die and electrical components external to it. The objective of fluid connectivity is to ensure the reliable transport of liquids and gases between the die and external fluid control units.

Electrical interconnects

Wire bonding

Wire bonding is unquestionably the most popular technique to electrically connect the die to the package. The free ends of a gold or aluminum wire form low-resistance (ohmic) contacts to aluminum bond pads on the die and to the package leads (terminals). Bonding gold wires tends to be easier than bonding aluminum wires.

Thermosonic gold bonding is a well-established technique in the integrated circuit industry, simultaneously combining the application of heat, pressure, and ultrasonic energy to the bond area. Ultrasound causes the wire to vibrate, producing localized frictional heating to aid in the bonding process. Typically, the gold wire forms a *ball bond* to the aluminum bond pad on the die, and a *stitch bond* to the package lead. The "ball bond" designation follows after the spherical shape of the wire end as it bonds to the aluminum. The stitch bond, in contrast, is a wedge-like connection as the wire is pressed into contact with the package lead (typically gold- or silver-plated). The temperature of the substrate is usually near 150° C, below the threshold of producing gold-aluminum intermetallic compounds that cause bonds to be brittle. One of these compounds is known as "purple plague" (Au_5Al_{12}), and is responsible for the formation of voids—the Kirkendall voids—by the diffusion of aluminum into gold. Thermosonic gold bonding can be automated using equipment commercially available from companies such as Kulicke and Soffa Industries, Inc., Willow Grove, Pennsylvania.

Bonding aluminum wires to aluminum bond pads is also achieved with ultrasonic energy, but without heating the substrate. In this case, a stitch bond works better than a ball bond, but the process tends to be slow. This makes bonding aluminum wires not as economically attractive as bonding gold wires. However, gold wires with diameters above 50 μm (2 mils) are difficult to obtain, which makes aluminum wires, available in diameters up to 560 μm (22 mils), the only solution for high-current applications (Table 6.3).

The thermosonic ball bond process begins with an electric discharge or spark to produce a ball at the exposed wire end. The tip—or capillary—of the wire-bonding tool descends onto the aluminum bond pad,

Table 6.3
Recommended Maximum Current in Gold and Aluminum Bond Wires

		Maximum current (A)	
Material	Diameter (μm)	Length < 1 mm	Length > 1 mm
Gold	25	0.95	0.65
	50	2.7	1.8
Aluminum	25	0.7	0.5
	50	2	1.4
	125	7.8	5.4
	200	15.7	10.9
	300	28.9	20
	380	40.4	27.9
	560	71.9	49.6

pressing the gold ball into bonding with the bond pad. Ultrasonic energy is simultaneously applied. The capillary then rises and the wire is fed out of it to form a loop as the tip is positioned over the package lead—the next bonding target. The capillary is lowered again, deforming the wire against the package lead into the shape of a wedge—the stitch bond. As the capillary rises, special clamps close onto the wire causing it to break immediately above the stitch bond. The size of the ball dictates a minimum in-line spacing of approximately 100 μm between adjacent bond pads on the die. This spacing decreases to 75 μm for stitch bonding (Figure 6.4).

The use of wire bonding occasionally runs into serious limitations in MEMS packaging. For instance, the applied ultrasonic energy, normally at a frequency between 50 and 100 kHz, may stimulate the oscillation of suspended mechanical microstructures. Unfortunately, most micromachined structures coincidentally have resonant frequencies in the same range, increasing the risk of structural failure during wire bonding.

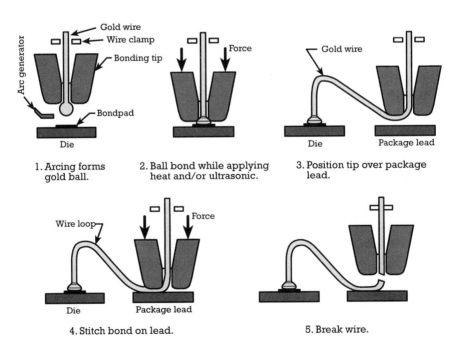

Figure 6.4 Illustration of the sequential steps in thermosonic ball and stitch bonding. The temperature of the die is typically near 150° C. Only the tip of the wire-bonding tool is shown [10].

Flip chip

Flip-chip bonding [11], as its name implies, involves bonding the die, top-face-down, on a package substrate. Electrical contacts are made by means of plated solder bumps between bond pads on the die and metal pads on the package substrate. The attachment is intimate with a relatively small spacing (50 ~ 200 μm) between the die and the package substrate. Unlike wire bonding which requires that bond pads are positioned on the periphery of the die to avoid crossing wires, flip-chip allows the placement of bond pads over the entire die (area arrays), resulting in a significant increase in density of input/output (I/O) connections—up to 700 simultaneous I/Os. In addition, the effective inductance of each interconnect is miniscule because of the short height of the solder bump. The inductance of a single solder bump is less than 0.05 nH, compared to 1 nH for a 125-μm-long and 25-μm-diameter wire. It becomes clear why the integrated circuit industry has adopted flip-chip for high-density, fast electronic circuits (Figure 6.5).

What makes flip-chip bonding attractive to the MEMS industry is its ability to closely package a number of distinct dice on a single package substrate with multiple levels of embedded electrical traces. For instance, one can use flip-chip bonding to electrically connect and package three accelerometer dice, a yaw-rate sensing die, and an electronic application-specific integrated circuit (ASIC) onto one ceramic substrate to build a fully self-contained navigation system. This type of hybrid packaging produces complex systems, though each individual component in itself may not be as complex. Clearly, a similar system can be built with wire bonding, but its area usage will not be as efficient, and its reliability may be questionable given the large number of gold wires within the package (note that each suspended gold wire is essentially an accelerometer, subject to deflections and potential shorting).

Additional fabrication steps are required to form the solder bumps over the die. A typical process involves the sputtering of a titanium layer over the bond pad metal (e.g., aluminum) to promote adhesion, followed by the sputtering of copper. Patterning and etching of the titanium and copper define a pedestal for the solder bump. A thicker layer of copper is then electroplated. Finally, the solder bump, typically a tin-lead alloy, is electroplated over the copper. Meanwhile, in a separate preparation process, solder paste is screen-printed on the package substrate in patterns corresponding to the landing sites of the solder bumps. Automated

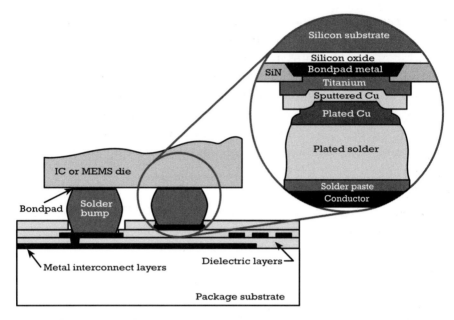

Figure 6.5 Flip-chip bonding with solder bumps.

pick-and-place machines position the die, top-face-down, and align the bond pads to the solder-paste pattern on the package substrate. Subsequent heating in an oven or under infrared radiation melts the solder into a columnar, smooth, and shiny bump. Surface tension of the molten solder is sufficient to correct for any slight misalignment during the die-positioning process. If desired, a final underfill step fills the void space between the die and the package substrate with epoxy. An optional silicone or parylene conformal coat protects the entire assembly.

Flip-chip may not be compatible with the packaging of MEMS that includes microstructures exposed to the open environment. For instance, there is a risk of damaging the thin diaphragm of a pressure sensor during a flip-chip process. In contrast, a capped device, such as the Bosch yaw-rate sensor (see Chapter 4), can take full advantage of flip-chip technology.

Microfluidic interconnects

All advances in electrical interconnect technology derive from the packaging requirements of the integrated circuit industry, but that is not the

The Box: Packaging for MEMS 221

case for fluid interconnects. These are required to package microfluidic devices such as micropumps and microvalves. No standards exist simply because the field remains in its infancy and few microfluidic devices are commercially available. Sadly, most microfluidic interconnect schemes remain at the level of manually inserting a capillary into a silicon cavity or via-hole, and sealing the assembly with silicone or epoxy (see, for example, the PCR thermal cycler in Chapter 5). These are suitable methods for laboratory experimentation, but will not meet the requirements of automated manufacturing (Figure 6.6).

Future fluid packaging schemes amenable to high-volume manufacturing would have to rely on simplified fluid interconnects. For example, fluid ports in a silicon die could be aligned directly to ports in a ceramic or metal manifold. The silicon die can be attached by any of the die-attach methods described earlier. Under such a scheme, it becomes possible to envision systems with fluid connectivity on one side of the die, and electrical connectivity on the opposite side. This would enhance long-term reliability by separating fluid flow from electrical wiring.

Researchers at Abbott Laboratories, Abbott Park, Illinois, demonstrated a hybrid packaging approach incorporating a complex manifold in an acrylic, for example Plexiglas™ [13]. These are large boards, many centimeters in size, with multiple levels of channels and access vias,

 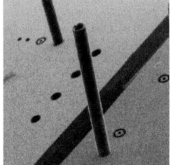

Figure 6.6 *Left:* Photograph of a fluid interconnect etched in silicon using deep reactive ion etching. Fluid flows through a central orifice leading into a channel embedded within the silicon substrate. The precise outer trench provides mechanical support to tightly hold a capillary in position. *Right:* Photograph of a capillary inserted into an intact fluid port. Courtesy of Lucas NovaSensor, Fremont, California [12].

all made in plastic. The channels are formed by laminating and bonding layers of thermoplastics into which trenches have been preformed. The plastic board becomes equivalent to a "fluid printed circuit board," onto which surface fluid components are attached and wired. These components need not necessarily be micromachined. For example, the board could hold a silicon pressure or flow sensor in proximity to a miniature solenoid valve. Much of the technology for fluid interconnects remains under development. New markets and applications will undoubtedly drive engineers to contrive innovative but economically justifiable solutions.

Types of packaging solutions

In its basic form, a package is a protective housing with an enclosure to hold one or more dice, which forms a complete microelectromechanical device or system. The package provides, when necessary, electrical and fluid connectivity between the dice and the external world.

In some cases, it is advantageous to provide a first level of packaging (chip- or die-level encapsulation) to the micromechanical structures and components [14]. This is of particular interest in applications where the surfaces of the microstructures do not need to be in direct exposure to liquids or gases. A top silicon cap attached, for example, by silicon-fusion bonding can maintain a hermetic seal and hold a vacuum while protecting the sensitive microstructures from damage during saw and assembly. A top cap also allows the use of plastic molding, ubiquitous in low-cost packaging solutions. In this method, molten plastic flows under high pressure, filling the inner cavity of a mold, and encapsulating a metal lead frame upon which the die or capped microstructure rests. For example, a crystalline silicon cap protects the sensing elements of the VTI Hamlin accelerometer (see Chapter 4) during molding of the plastic package over the die. Fixed to ground potential, the cap also becomes an effective shield against electromagnetic interference [15].

There are three general categories of widely adopted packaging approaches in MEMS. They are ceramic, metal, and plastic, each with its own merits and limitations. For instance, plastic is a low-cost and often small size (surface-mount) solution, but it is inadequate for harsh environments. The asking price for a plastic-packaged pressure or acceleration

The Box: Packaging for MEMS

sensor is frequently below $5. In contrast, a similar sensor packaged in a hermetic metal housing may cost well over $30. It is not surprising, therefore, that packaging is what frequently determines economic competitiveness (Table 6.4).

Ceramic packaging

Ceramics are hard and brittle materials made by shaping a nonmetallic mineral, then firing at a high temperature for densification. The vast majority of ceramics are electrical insulators, and often good thermal conductors, also. Ease of shaping along with reliability and attractive material properties (Table 6.5) (e.g., electrical insulator, hermetic sealing) have made ceramics a mainstay in electronic packaging. They are widely used in multichip modules (MCM) [16] and advanced electronic packages, such as ball grid arrays (BGA) [17]. These same characteristics have extended the utility of ceramics to the packaging of MEMS—many commercially available micromachined sensors use some form of ceramic packaging. Ceramics are completely customizable and allow the formation of through-ports and manifolds for the packaging of fluid-based MEMS. But ceramics usually suffer from shrinkage ($\sim$ 13% in the horizontal direction and $\sim$ 15% in the vertical direction) during firing. Compared with plastic packaging, they are significantly more expensive.

Aluminas (Al_2O_3) are by far the most common of all ceramics, having been used over the centuries in porcelain and fine dinnerware. Aluminum nitride (AlN) and beryllia (BeO) have superior material properties (e.g., better thermal conductivity), but the latter is very toxic. Aluminum nitride substrates, in particular, tend to be costly because of required complex processing due to the difficulty of sintering the material.

A ceramic package is made of laminates, each formed and patterned separately, then brought together and cofired (sintered) at an elevated temperature, typically between 1500° and 1600° C (Figure 6.7). Recent advances have led to low-temperature cofired ceramics (LTCC), such as the Dupont 951 Green Tape™, with sintering temperatures near 800° C. Powders are first mixed together with special additives and extruded under a knife edge to form a thin laminate sheet. This "green" unfired soft tape, approximately 0.1- to 0.3-mm-thick, is peeled from the supporting table, then cut and punched using precise machining tools. Patterns of electrical interconnects are screen-printed on each sheet using a slurry of

Table 6.4
Table Illustrating the Diversity of MEMS Packaging Requirements

	Electrical contacts	Fluid[1] ports	Media contact	Transparent window	Hermetic sealing	Stress isolation	Heat sinking	Thermal isolation	Calibration & compensation	Types of packaging[2]
Sensors										
Pressure	Yes	Yes	Yes	No	Possibly	Yes	No	No	Yes	P, M, C
Flow	Yes	Yes	Yes	No	No	No	No	Yes	Yes	P, M, C
Acceleration	Yes	No	No	No	Yes	Possibly	No	No	Yes	P, M, C
Yaw-rate	Yes	No	No	No	Yes	Possibly	No	No	Yes	P, M, C
Microphone	Yes	Yes	Yes	No	No	No	No	No	Yes	P, M, C
Hydrophone	Yes	Yes	Yes	No	Possibly	No	No	No	Yes	M, C
Actuators										
Optical switch	Yes	No	No	Yes	Yes	No	No	No	Yes	C
Display	Yes	No	No	Yes	Yes	No	Possibly	Possibly	No	C
Valve	Yes	Yes	Yes	No	No	No	Possibly	Possibly	Possibly	M, C
Pump	Yes	Yes	Yes	No	No	No	Possibly	No	Possibly	M, C
PCR thermal cycler	Yes	Yes	Yes	Possibly	No	No	Possibly	Yes	No	P, M, C
Electrophoresis	Yes	Yes	Yes	Yes	No	No	No	No	No	P, M, C
Passive										
Nozzles	No	Yes	Yes	No	No	No	No	No	No	P, M, C
Fluid mixer	No	Yes	Yes	Possibly	No	No	No	No	No	P, M, C
Fluid amplifier	No	Yes	Yes	No	No	No	No	No	Possibly	M, C

[1] Fluid includes liquid or gas.   [2] P: Plastic, M: Metal, C: Ceramic.

Table 6.5
Material Properties of Some Notable Ceramics Compared with Silicon

Ceramic	Relative permittivity	Thermal conductivity (W/m·K)	Thermal expansion (10^{-6}/°C)	Density (g/cm^{-3})
Alumina (Al_2O_3)	9.7	40	7.2	4
Aluminum Nitride (AlN)	10	150	2.7	3.2
Beryllia (BeO)	6.8	300	7	2.9
Borosilicate glass	3.7	2	3.2	2.1
Silicon	11.8	157	2.6	2.4

tungsten powder. This process also fills via holes with metal. Vias left unfilled with tungsten can be later used as fluid or pressure access ports through the ceramic. Several "green" sheets are aligned and press-laminated together, then cofired at an elevated temperature in a reducing atmosphere to sinter the laminate stack into a monolithic body. A typical integrated circuit package consists of three laminates, but as many as sixteen may be simultaneously cofired, naturally at a higher material cost. An appropriate metal finish is then applied to the tungsten, followed by plating of nickel. If necessary, pins or leads are brazed to the package. The leads are typically made of ASTM F-15 alloy ($Fe_{52}Ni_{29}Co_{18}$), also known as Kovar®. The brazing material is often a silver-copper eutectic alloy. A final electroless gold-plating step ensures that wires can be bonded to the leads. A ball grid array (BGA) ceramic package has no pins brazed, rather it has arrays of solder balls connected to electrical feed-throughs. One attractive feature of ceramic is the ability to screen print on its surface a network of thick-film resistors that can be later trimmed with a laser for sensor calibration.

Whether custom or standard, a ceramic package often consists of a base or a header onto which one or many dice are attached by adhesives or solder. Wire bonding is suitable for electrical interconnects. Flip-chip bonding to a pattern of metal contacts on the ceramic works equally well. The final step after mounting the die on the base and providing suitable electrical interconnects involves capping and sealing the assembly with a lid, whose shape and properties are determined by the final application. For instance, the lid must be transparent for optical MEMS, or must hermetically seal a vacuum, as is the case for the infrared bolometer from

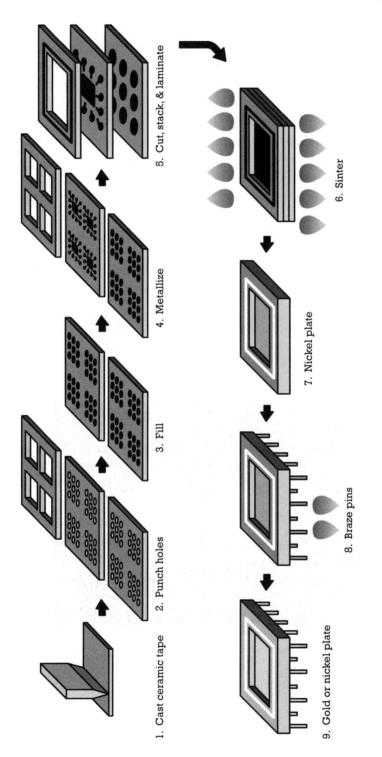

Figure 6.7 Process flow for the fabrication of a cofired, laminated ceramic package with electrical pins and access ports. Courtesy of the Coors Electronic Package Company, Chattanooga, Tennessee.

The Box: Packaging for MEMS   227

Honeywell or the DMD™ from Texas Instruments (see Chapter 4). In contrast, a plastic cover provides a cost-effective solution for low-cost devices. For example, disposable blood pressure sensors used for arterial line measurement in intensive care units, are protected by a plastic cover that includes an access opening for pressure [18]. A special gel dispensed inside this opening provides limited protection (particularly against biological solutions) to the device, while permitting the transmission of pressure to the sensitive silicon membrane (Figure 6.8).

Ceramic packaging of optical MEMS can be complex and costly. This is certainly true for DMD™ packages, which have undergone a continuous evolution from their early application in Airline Ticket Boarding (ATB) printers to today's high-resolution display arrays [19]. The DMD™ type-A package for SVGA displays consists of a 114-pin alumina (Al_2O_3) ceramic header (base), with metalization for electrical interconnects, and a Cu-Ag brazed Kovar® seal ring. Wire bonds establish electrical connectivity between the die and metal traces on the ceramic header. A transparent window, consisting of a polished Corning 7056 glass fused to a stamped gold-nickel-plated Kovar® frame, covers the assembly. Resistance seam welding of the seal ring on the ceramic base to the Kovar®

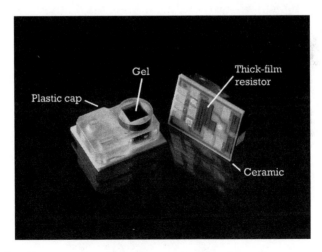

Figure 6.8 Photograph of a disposable blood pressure sensor for arterial-line measurement in intensive care units. The die sits on a ceramic substrate and is covered with a plastic cap that includes an access opening for pressure. A special gel dispensed inside the opening protects the silicon device while permitting the transmission of pressure. Courtesy of Lucas NovaSensor, Fremont, California [18].

glass frame provides a permanent hermetic seal. Two zeolite getter strips attached to the inside of the glass window ensure long-term desiccation. The particular choice of metal and glass window materials minimizes the mismatch in coefficients of thermal expansion (4×10^{-6} and 5×10^{-6} per °C for Kovar® and Corning 7056, respectively), and reduces stresses during the high temperature ($\sim 1000°$ C) metal-to-glass fusing process. Antireflective coatings applied to both sides of the glass window reduce reflections to less than 0.5%. A heat sink attached to the backside of the ceramic package by means of adhesives keeps the temperature of the DMD™ within tolerable limits (Figure 6.9).

Metal packaging

In the early days of the integrated circuit industry, the number of transistors on a single chip, and the corresponding pin count (number of input/output connections) were few. Metal packages were practical because they were robust and easy to assemble. The standard family of TO-type (Transistor Outline) packages grew to cover a wide range of shapes, but all accommodated fewer than 10 electrical pins. The semiconductor industry abandoned the TO-packages in favor of plastic and ceramic packaging, as the density of transistors grew exponentially and the required pin count increased correspondingly. Today, TO-type packages remain in use for a few applications, in particular high-

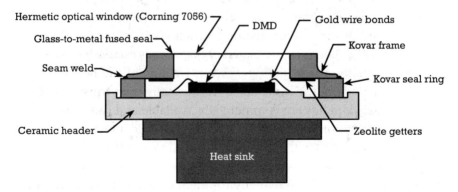

Figure 6.9 Illustration of the DMD™ type-A ceramic package. The assembly includes a hermetically sealed optical window for high-resolution projection display [19].

The Box: Packaging for MEMS

power discrete devices and high-voltage linear circuits (e.g., operational amplifiers).

Metal packages are attractive to MEMS for the same reasons the integrated circuit industry adopted the technology over 30 years ago. They satisfy the pin-count requirements of most MEMS applications; they can be prototyped in small volumes with rather short turnaround periods; and they are hermetic when sealed. But a major drawback is the relatively large expense of metal headers and caps; they cost a few dollars per assembled unit, at least ten-times higher than an equivalent plastic package. Early prototypes of the ADXL family of accelerometers from Analog Devices (see Chapter 4) were available in TO-type hermetic metal packages. However, pressure to reduce manufacturing costs has led the company to adopt a standard plastic dual-in-line (DIP) solution, and to establish first-level packaging (at the die level) using proprietary chip-encapsulation methods.

A metal hermetic package, including the familiar TO-8 type, is frequently made of ASTM F-15 alloy (Kovar®), though stainless steel is also common. A sheet of metal is first formed into a header or a tub-like housing. Holes are then punched, either through the bottom for plug-in packages, or the sides for flat packages. An oxide is then grown over the package housing. Metal leads are placed through the holes and beads of borosilicate glass, such as Corning 7052 glass, and placed over the leads. Fusing of the glass to metal at a temperature above the melting temperature of glass ($\sim 500°$ C) produces a hermetic metal-to-glass seal. Etching the metal oxides reveals a fresh alloy surface, which is then plated with either nickel or gold—both of which allow wire bonding and soldering. The die containing the micromachined structures is mounted directly on the header. Wire bonds to the plated package leads establish electrical connectivity. Finally, the soldering or seam welding of the header to a lid, most often made of the same alloy, hermetically seals the assembly. Standard headers and lids are commercially available and can be readily modified in conventional machine shops. For instance, metal tubes can be brazed to drilled ports in the header and the lid to provide access to fluids in pressure and flow sensors, and microvalves (Figure 6.10).

Packaging solutions for harsh environments, namely those found in heavy industries and aerospace, can be complex and costly. The custom requirements of the application coupled with the lack of high-volume market demand, has turned packaging for harsh environments into a

Figure 6.10 Modified by brazing two tubes to the header and the cap, the TO-8 metal can becomes suitable for packaging fluidic microdevices such as this microvalve from Redwood Microsystems. Courtesy of A. Henning, Redwood Microsystems, Menlo Park, California.

niche art. One particularly interesting design is the metal packaging of media-isolated pressure sensors for operation in heavily industrial environments. The design immerses the silicon pressure sensor within an oilfilled stainless-steel cavity that is sealed with a thin, stainless steel diaphragm. The silicon pressure sensor measures pressure transmitted via the steel diaphragm and through the oil. The robust steel package offers hermetic protection of the sensing die and the wire bonds against adverse environmental conditions.

Each stainless-steel package is individually machined to produce a cavity. Electrical pins are glass-fired in holes through the steel housing. The die is attached inside the cavity, and wire bonding to the electrical pins is completed. Welding of a stainless-steel diaphragm seals the topside of the assembly. Oil filling of the cavity occurs through a small port at the bottom, which is later plugged and sealed by welding a ball (Figure 6.11).

Molded plastic packaging

Unlike metal or ceramic packages, molded plastic packages are not hermetic. Yet, they dominate in the packaging of integrated circuits because they are cost-effective solutions (costing on average a few pennies or less

The Box: Packaging for MEMS

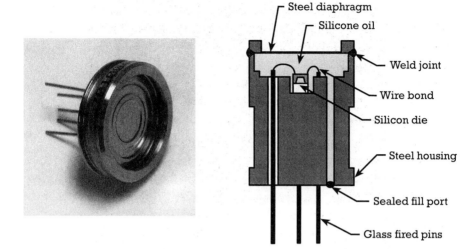

Figure 6.11 Photograph (left) and cross-sectional schematic (right) of a pressure sensor mounted inside an oil-filled, stainless steel package. Pressure is transmitted via the stainless-steel diaphragm and through the oil to the silicon sensor. Courtesy of Lucas NovaSensor, Fremont, California.

per electrical pin). Advances in plastic packaging have further improved reliability to high levels. Today's failure rates in plastic-packaged logic and linear integrated circuits are less than one failure in every ten billion hours of operation [20].

There are two general approaches to plastic packaging: Postmolding and premolding. In the first approach, the plastic housing is molded after the die is attached to a "lead frame" (a supporting metal sheet). The process subjects the die and the wire bonds to the harsh molding environment. In premolding, the die is attached to a lead frame over which plastic was previously molded. It is attractive in situations where the risk of damaging the die is high, or if openings through the plastic are necessary (e.g., for pressure or flow sensors). However, it tends to be more expensive than postmolding.

The metal lead frame in either approach is an etched or stamped metal sheet consisting of a central platform (paddle) and metal leads supported by an outer frame. The leads provide electrical connectivity and emanate from the paddle in the shape of a fan. The metal is typically a copper alloy or Alloy-42 ($Ni_{42}Fe_{58}$); the latter has a coefficient of thermal expansion (4.3×10^{-6} per °C) that closely matches that of silicon.

In postmolded plastic packaging, the lead frame is spot-plated with gold or silver on the paddle and the lead tips to improve wire bonding. The die is then attached with adhesive or eutectic solder. Wires are bonded between the die and the lead tips. Plastic molding encapsulates the die and lead frame assembly, but leaves the outer edges of the leads exposed. These leads are later plated with tin or tin-lead, to improve wetting during soldering to printed circuit boards. Finally, the outer frame is broken off and the leads are formed into an S-shape (Figure 6.12).

The sequence of process steps differs for premolded plastic packages. First, a plastic body is molded onto a metal lead frame. The molded thermosetting plastic polymer encapsulates the entire lead frame, with the exception of the paddle and the outer edges of the leads. Deflashing of the package removes any undesirable or residual plastic on the die-bonding areas. The molded body may contain ports or openings which later may be used to admit a fluid (e.g., for pressure or flow sensing). The lead frame is spot-plated with gold or silver to improve wire bonding and soldering. At this point, the die is attached and wire-bonded to the lead frame. A protective encapsulant, such as RTV or silicone gel, is then dispensed over the die and wire bonds. Finally, a premolded plastic cap is attached, using

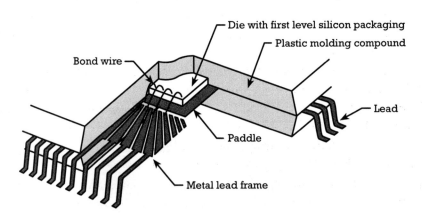

Figure 6.12 Schematic showing a sectional view of a postmolded plastic package. The die is first mounted on a center platform (the paddle) and wires bonded to adjacent electrical leads. The paddle and the leads form a metal "lead frame" over which the plastic is molded. A MEMS die should include a first level of packaging (e.g., a bonded silicon cap) as protection against the harsh effects of the molding process. This particular illustration is of a plastic quad-flat pack (QFP), with electrical leads along its entire outer periphery.

The Box: Packaging for MEMS

an adhesive. If necessary, the cap itself may also contain a fluid access port (Figures 6.13 and 6.14).

The molding process is a harsh process which involves melting the thermosetting plastic at approximately 175° C, then flowing it under relatively high pressure (~ 6 MPa) into the mold cavity before it is allowed to cool. The plastic material is frequently an epoxy. Novolac epoxies are preferred because of their improved resistance to heat. The temperature cycle gives rise to severe thermal stresses, due to the mismatch in coefficients of thermal expansion between the plastic, the lead frame, and the die. These stresses may damage the die, or cause localized delamination of the plastic. The material properties of the plastic, and especially its coefficient of thermal expansion, are carefully adjusted by the introduction of additives to the epoxy. Fillers such as glass, silica, or alumina powder make up 65 to 70% of the weight of the final product, and help tailor its coefficient of thermal expansion as well as its thermal conductivity. In addition, mold-release agents (e.g., synthetic or natural wax) are introduced to promote the release of the plastic part from the mold. Flame-retardant materials, typically brominated epoxy or antimony trioxide, are also added to meet industry flammability standards. Carbon and other organic dyes give the plastic its commonly black appearance that is necessary for laser marking.

Plastic packaging for integrated circuits (IC) is governed by standards set forth by the Electronics Industries Association (EIA), the Joint Electron Device Engineering Council (JEDEC), and the Electronics Industry

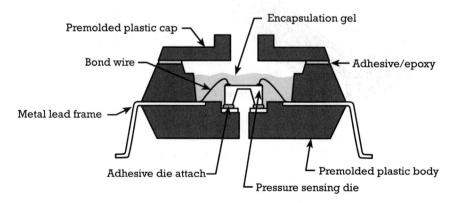

Figure 6.13 Illustration of a premolded plastic package [21]. Adapting it to pressure sensors involves incorporating fluid ports in the premolded plastic housing and the cap.

Figure 6.14 Photograph of the NovaSensor NPP-301, a premolded plastic-surface-mount (SOIC-type) absolute pressure sensor. Courtesy of Lucas NovaSensor, Fremont, California.

Table 6.6
Selected Standard Molded-plastic Packages for Integrated Circuits [22]*

	Type	Abbreviation	Pin count	Description
Surface-mount	Small outline IC	SOIC	min. 8, max. 28	Small package with leads on two sides.
	Thin small outline package	TSOP	min. 26, max. 70	Thin version of the SOIC.
	Small outline J-lead	SOJ	min. 24, max. 32	Same as SOIC, but with leads bent in J-shape.
	Plastic leaded chip carrier	PLCC	min. 18, max. 84	J-shaped leads on 4 sides.
	Thin quad flat pack	TQFP	min. 32, max. 256	Wide but thin package with leads on 4 sides.
Through hole-mount	Dual in-line	DIP	min. 8, max. 64	Two in-line row of leads.
	Single in-line	SIP	min. 11, max. 40	One in-line row of leads.
	Zigzag in-line	ZIP	min. 16, max. 40	Two rows with staggered leads.
	Quad in-line package	QUIP	min. 16, max. 64	Four in-line rows of leads. Leads are staggered.

* Surface-mount devices are generally thinner than through-hole-mount packages, and accommodate a smaller spacing between adjacent leads (pins).

Association of Japan (EIAJ) (Table 6.6). While plastic packaging for MEMS is not governed by any standards yet, it often uses standard or slightly modified IC plastic packages. The development of new plastic packaging technologies for MEMS will most likely remain in the distant future, because of the prohibitive associated costs.

Summary

Packaging of MEMS is an art rather than a science. The diversity of MEMS applications places a significant burden on packaging. Standards do not exist in MEMS packaging; rather the industry has opted to borrow standards and methods from the integrated circuit industry, and modify them as necessary. This chapter reviewed the basic considerations of MEMS packaging and introduced three widely accepted packaging approaches: ceramic, metal, and plastic.

References

[1] Lau, J. H., et al., *Electronic Packaging: Design, Materials, Process, and Reliability*, New York, NY: McGraw-Hill, 1998, pp. 111–193.

[2] Beach, W. F., T. M. Austin, and R. Olson, "Parylene Coatings." In *Electronic Materials Handbook: Volume 1, Packaging*, pp. 789–801, M. L. Minges, C. A. Dostal, and M. S. Woods (eds.), Materials Park, OH: ASM International, 1989.

[3] Monk, D. J., et al., "Media Compatible Packaging and Environmental Testing of Barrier Coating Encapsulated Silicon Pressure Sensors," *Tech. Digest Solid-State Sensor and Actuator Workshop*, Hilton Head Island, SC, June 3–6, 1996, pp. 36–41.

[4] Flannery, A. F., et al., "PECVD Silicon Carbide for Micromachined Transducers," *Proc. 1997 Int. Conf. on Solid-State Sensors and Actuators*, Chicago, IL, June 16–19, 1997, Vol. 1, pp. 217–220.

[5] Application notes AN 840, AN 935, AN 1097, AN 1315, AN 1318, Motorola, Inc., 1999, http://www.mot_sps.com/lit/index/Applications.html.

[6] Sergent, J. E., "The Hybrid Microelectronics Technology." In *Electronic Packaging & Interconnection Handbook*, 2nd ed., pp. 7.10–7.26, C. A. Harper (ed.), New York, NY: McGraw-Hill, 1997.

[7] Hwang, J. S., "Solder Technologies for Electronic Packaging." In *Electronic Packaging & Interconnection Handbook*, 2nd ed., pp. 5.4–5.20, C. A. Harper (ed.), New York, NY: McGraw-Hill, 1997.

[8] Striny, K. M., "Assembly Techniques and Packaging of VLSI Devices." In *VLSI Technology*, 2nd ed., pp. 566–611, S. M. Sze (ed.), New York, NY: McGraw-Hill, 1988.

[9] DowCorning Corporation, Midland, MI 48686–0994, (517) 496–6000, http://www.dowcorning.com/html/industries/electronics/index.html.

[10] *Bonding Handbook and General Catalog*, Kulicke and Soffa Industries, Inc., Willow Grove, PA, 1990.

[11] *Flip Chip Technologies*, J. H. Lau (ed.), New York, NY: McGraw-Hill, 1996.

[12] Jaeggi, D., et al.,, "Novel Interconnection Technologies for Integrated Microfluidic Systems," *Tech. Digest Solid-State Sensor and Actuator Workshop*, Hilton Head Island, SC, June 8–11, 1998, pp. 112–115.

[13] VerLee, D. ,et al., "Fluid Circuit Technology: Integrated Interconnect Technology for Miniature Fluidic Devices," *Tech. Digest Solid-State Sensor and Actuator Workshop*, Hilton Head Island, SC, June 3–6, 1996, pp. 9–14.

[14] Eddy, D. S. and D. R. Sparks, "Application of MEMS Technology in Automotive Sensors and Actuators," in *Integrated Sensors, Microactuators, & Microsystems (MEMS)*, pp. 1750–1751, K. D. Wise (ed.), Proceedings of the IEEE, Vol. 86, No. 8, Aug. 1998.

[15] US Patent #5,545,912 (Aug. 13, 1996).

[16] Ginsberg, G. L., and D. P. Schnorr, *Multichip Modules and Related Technologies*, New York, NY: McGraw-Hill, 1994.

[17] *Ball Grid Array Technology*, J. H. Lau (ed.), New York, NY: McGraw-Hill, 1995.

[18] NPC-107 data sheet, Lucas NovaSensor, 1055 Mission Court, Fremont, California 94539, http://www.novasensor.com.

[19] Faris, J., and T. Kocian, "DMD™ Packages—Evolution and Strategy," *TI Technical Journal*, July-Sept. 1998, pp. 87–94.

[20] Bonner, J. K., "Surface Mount Technology." In *Electronic Packaging & Interconnection Handbook*, 2nd ed., pp. 9.50–9.51, C. A. Harper (ed.), New York, NY: McGraw-Hill, 1997.

[21] Cohn, C., and M. T. Shih, "Packaging and Interconnection of Integrated Circuits." In *Electronic Packaging & Interconnection Handbook*, 2nd ed., pp. 6.14–6.17, C. A. Harper (ed.), New York, NY: McGraw-Hill, 1997.

[22] *ibid*, pp. 6.46–6.56.

Selected bibliography

Electronic Packaging & Interconnection Handbook, 2nd ed., C. A. Harper (ed.), New York, NY: McGraw-Hill, 1997.

Harper, C. A., and M. B. Miller, *Electronic Packaging, Microelectronics and Interconnection Dictionary*, New York, NY: McGraw-Hill, 1993.

Lau, J. H., et al., *Electronic Packaging: Design, Materials, Process, and Reliability*, New York, NY: McGraw-Hill, 1998.

Striny, K. M., "Assembly Techniques and Packaging of VLSI Devices." In *VLSI Technology*, 2nd ed., pp. 566–611, S. M. Sze (ed.), New York, NY: McGraw-Hill, 1988.

Glossary

Action potential A temporary change in the electrical voltage across the cellular membrane of a nerve or muscle cell. Action potentials occur when the cell is stimulated, especially by a nerve impulse. They form the mechanism by which sensory and motor functions are transmitted across the nervous system.

Amorphous silicon Silicon lacking a preferred crystalline orientation, typically consisting of extremely fine grains each measuring a few nanometers in size.

Amplification In biochemistry, it is the process of making a large number of identical copies of a DNA fragment. In electronics, it is the process of increasing the magnitude of an electrical voltage.

Anodic bonding A process to bond silicon to glass, specifically Pyrex® or equivalent.

Application-specific integrated circuit (ASIC) An electronic integrated circuit with a functionality designed specifically for one particular application; for example, the detection of minute changes in capacitance and conversion into an output voltage.

Bandpass *see* filter.

Bandwidth The extent of the frequency response of a linear system. It is numerically defined as the difference between two corner frequencies where the system gain is 3 dB below the maximal gain. An input signal with a frequency content below or above the corner frequencies is severely attenuated by the system.

Bimetallic actuation The resulting motion when a stack of two materials having dissimilar coefficients of thermal expansion is heated. One material expands more than the other, giving rise to bending stresses. The amount of bending is proportional to the temperature of the stack and the difference in coefficients of thermal expansion.

BGA An acronym for ball grid array. A type of advanced ceramic package for integrated circuits consisting of arrays of solder balls instead of electrical pins.

Bond pad A metal area on a die or wafer to which a gold or aluminum wire is bonded. The wire and bond pad provide electrical connectivity between electrical components on the die and electronic circuitry external to the die.

Brownian noise *see* noise.

Bulk micromachining A class of micromachining processes that yield micromechanical structures with thicknesses in the tens or hundreds of micrometers. Very often, it also refers to micromechanical structures formed using wet anisotropic etch solutions, such as potassium hydroxide.

Capacitor microphone, *syn.* **condenser microphone** Microphone in which acoustic energy is converted to electricity by varying the capacitance of a deflecting membrane suspended over a fixed back plate.

CERDIP An acronym for a ceramic dual-in-line package for integrated circuits. It consists of a rectangular pressed ceramic body with pins on two opposite sides. A ceramic cap is glass sealed to the body.

CERQUAD An acronym for a ceramic quad-flat pack. Similar to the CERDIP but it has pins on all sides.

Check valve A valve that permits fluid flow in one direction only.

Glossary

Chemical vapor deposition (CVD)   A process based on the principle of initiating a chemical reaction in a vacuum chamber, resulting in the deposition of a reacted species on a heated substrate. Materials that can be deposited by CVD include polysilicon, silicon oxide, and silicon nitride.

CMOS   An acronym for complementary metal oxide semiconductor. A class of electronic devices made of silicon and associated fabrication processes common in integrated circuits.

Coefficient of thermal expansion (CTE)   The rate of change in length of an object as a function of temperature. In general, $CTE = (\Delta L/L)/\Delta T$, where $(\Delta L/L)$ is the fractional change in length corresponding to a ΔT change in temperature. It is measured in inverse units of temperature ($/°C$).

Complementarity   In biochemistry, it is the specific affinity for binding between the purines (adenine and guanine) with pyrimidines (thymine and cytosine). *See* nucleotide.

Condenser microphone   *see* capacitor microphone.

Coriolis effect   Physical effect responsible for the deflection of objects moving on the surface of a rotating body such as Earth.

Corner frequency   *see* filter.

Decibel, *abbr.* **dB**   A unit to measure the relative difference in intensity of a physical, electrical, or acoustical signal. It is defined as 20 times the 10-base logarithm of the intensity ratio. For example, 40 dB is equivalent to a ratio of 100 between the highest and lowest value in the range.

Degeneracy   *see* frequency degeneracy.

Die   Also chip, a common term in microfabrication technology indicating a small piece of semiconductor or glass cut or diced from a much larger wafer.

Diffusion   In semiconductor fabrication, it is the process to controllably spread or diffuse impurity dopant atoms in silicon or a semiconductor. Diffused resistors are resistors made of one type of doping (e.g., *p*-type) and embedded in silicon with a background doping of the opposite type (e.g., *n*-type).

DIP   An acronym for a dual-in-line type of package. Made of ceramic or plastic, it is rectangular in shape with pins (leads) on its two long sides.

Dipole   Also electric dipole, it is the electric field created by two charges of equal magnitude but of opposite polarity, and separated by a small distance.

Doping   Also known as impurity doping. A process of introducing into a semiconductor material impurities or foreign atoms—dopants—in relatively dilute concentrations ($10^{13} \sim 10^{20}$ cm^{-3}). The impurities alter the electrical properties of the semiconductor by adding electrons or holes (carriers). At or near room temperature, the carrier concentration is largely equal to the dopant concentration. If the doping is n-type, then there is an excess of electrons. Conversely, a p-type material has an excess of holes. The material is electrically more conductive at higher doping levels. A p-type region in direct contact with n-type forms a p-n diode, which passes current only in one direction: from the p-type to the n-type. Arsenic and phosphorous are common n-type dopants in silicon; boron is a p-type dopant in silicon.

Duality   A generally abstract concept that pairs equivalent parameters from distinct physical systems on the basis of energy arguments. Duality is frequently invoked between mechanical, thermal, and electrical systems. For example, a spring in a mechanical system is dual to a capacitor because they both store potential energy. In general, duality pairs mass to inductance, spring constant to the inverse of capacitance, coefficient of viscous damping to resistance, mechanical displacement to charge, velocity to electrical current, and applied force to applied voltage.

EEPROM   An acronym for electrically erasable and programmable read-only memory, a type of read-only electronic memory that can be erased and re-programmed using high voltage electrical pulses.

Electret microphone   Capacitor microphone in which a permanently polarized dielectric (an electret) produces a persistent charge and a polarizing voltage across the capacitor plates.

Electromagnetic interference (EMI)   Undesirable interference with electronic signals of electromagnetic nature. Sources of EMI include solar eruptions, radio signals, and nuclear explosions.

Electrophoresis   In chemistry, it is the migration of charged or polar molecules in colloidal suspension through a solution under the effect of an externally applied electric field. It is useful for the separation of dissimilar molecules and analysis of their molecular structure based on their rate of movement.

Epitaxy   Chemical process to grow a thin crystalline layer on top of a crystalline substrate. The grown layer generally has the same crystalline orientation as the substrate.

Eutectic point   At their eutectic point, alloys have identical solidus and liquidus temperatures. The melting temperature of a eutectic alloy is lower than that of any other alloy composed of the same constituents in different proportions. *See* liquidus temperature.

Filter   In electronics, a circuit that selectively blocks the transmission of certain frequencies. The transition frequencies defining the bands of transmission are known as "corner frequencies." A low-pass filter blocks high frequencies, but permits the transmission of low frequencies. A high-pass filter performs the opposite function. A bandpass filter allows the transmission of frequencies in a mid-band range, but blocks the transmission of frequencies above or below this band—outside of the corner frequencies.

Foundry   A service facility capable of prototyping and fabricating semiconductor circuits or MEMS. Foundry services typically offer a set of standard fabrication processes. A few provide custom design services.

Frequency degeneracy   The situation when two or more resonant modes oscillate at exactly the same frequency. When a number of identical oscillators are coupled with each other, their frequencies become degenerate. The coupling generally lifts this degeneracy by splitting the frequencies apart. The amount of separation depends on the strength of the coupling.

Helmholtz cavity   In acoustics, also known as a Helmholtz resonator, it is a hollow air-filled cavity having an inlet opening for sound and an outlet. The cavity is an acoustic oscillator with a characteristic resonant frequency determined by the air volume, and the geometry of the inlet and outlet ports. It is commonly used in acoustics for frequency tuning.

Hole   In physics, it is a vacant position in a semiconductor left by the absence of an electron. The concept is analogous to a bubble in water left by the absence of liquid. A hole is a carrier of positive electric charge, and participates in electric conduction.

Hybridization   In biochemistry, it is the process when two DNA strands having complementary sequences of nucleotides match up and bind with each other.

Impedance   A measure of the total resistance to electrical current flow. In acoustics, it is a measure of the total resistance to the propagation of acoustic pressure waves through a medium.

Insertion loss   In a linear system, such as a filter, it is the attenuation, measured in dB, of an input signal with a frequency content within the system bandwidth. Ideally, it is zero.

Ion implantation   A high-energy process capable of embedding impurity dopant atoms within the surface of a semiconductor substrate. It is usually followed by a high-temperature diffusion or anneal step. Implantation is useful in the doping of piezoresistors, embedded electrical interconnects in a silicon substrate, and thin polysilicon films.

Liquidus temperature   In metallurgy, the phase state of an alloy changes with temperature, pressure, and mole percentage of its constituents. At a constant pressure, there are three distinct regions in the phase diagram. At low temperatures, the alloy is a solid. At high temperatures, it is a liquid. A third intermediate region defines a plastic-like, mixed-liquid and solid state. The dividing line between the liquid state and the plastic-like state is the "liquidus" line. The dividing line between the solid state and the plastic-like state is the "solidus" line. For each molar composition, there is a liquidus and a solidus temperature. At the eutectic composition, the two temperatures are identical; in other words, the two lines coalesce and the plastic-like phase vanishes.

Lithography   A process common in microfabrication for delineating a pattern image in a photosensitive polymer. The polymer, or photoresist, can then be used as a masking layer to transfer the pattern into the underlying substrate.

Lorentz force　　In physics, it is the force on a current-conducting element inside a magnetic field. The force is equal to the current multiplied by the strength of the magnetic field and the length of the conductor.

LPCVD　　An acronym for low pressure chemical vapor deposition.

MESFET　　An acronym for metal semiconductor field effect transistor. A type of electronic transistor useful for operation at very high frequencies. It is very common in electronic circuits made of gallium arsenide (GaAs).

Microelectromechanical systems (MEMS)　　A generic descriptive term, common in the United States, for a broad technology having the objective of miniaturizing complex systems by integrating a diverse set of functions into a small package, or often a single die.

Micromachining　　A term describing the set of design and fabrication tools for the machining of microstructures and very small mechanical features in a substrate frequently made of silicon.

Multichip modules (MCM)　　A type of high-density packaging approach common in the integrated circuit industry that involves electrically connecting a number of dice on the same substrate.

n-type doping　　*see* doping.

Noise　　A random disturbance in an electrical or mechanical signal. It frequently determines the resolution of a sensor. Noise originates from a multitude of sources. The most common is electrical interference noise and may be filtered. Thermal—or Brownian—and $1/f$ noise are fundamental physical entities. Thermal noise originates from physical mechanisms where energy is converted to heat; for example, electrical resistance or mechanical friction. It is white in nature, meaning its spectral energy density is constant over frequency. Thermal noise increases with temperature. In contrast, $1/f$ noise, as the name implies, has a spectral energy density that decreases at higher frequencies. It is common in electronic circuits, and originates from crystal imperfections that momentarily trap electrons (hence the frequency dependence). The corner frequency in a noise spectrum is the frequency where $1/f$ and thermal noise are equal. A common measure of electrical noise is the magnitude of the spectral energy density at a particular frequency, given in $V/\sqrt{Hz}$.

Nucleotide The building block of deoxyribonucleic acid (DNA). It can be adenine (A), cytosine (C), guanine (G), or thymine (T). The sequence of nucleotides in DNA defines the genetic code.

Numerical aperture In optics, it is the sine of the angle that a ray of light makes at the image with the optic axis of a lens or lens system. It is the inverse of the *f*-number and a measure of the aperture size. The numerical aperture is always less than one.

Oxidation Chemical process by which the atoms of an element lose electrons. In an aqueous solution, neutral atoms become positive ions.

PECVD An acronym of plasma enhanced chemical vapor deposition.

Phase quadrature *see* quadrature.

Piezoelectricity The property exhibited by a class of materials to develop a voltage in response to applied mechanical stress or pressure. Conversely, an externally applied electrical voltage strains and deforms the material.

Piezoresistivity The property of a certain class of materials, including impurity-doped silicon, to change their electrical resistivity in response to mechanical stress.

Plasma etching A class of etch processes capable of selectively removing material, including silicon, by chemical reaction with one or more gases. The reactive gases are ionized in a plasma inside a vacuum chamber by means of electrical or electromagnetic energy. A plasma is an electrically neutral, highly ionized gas composed of ions, electrons, and neutral particles.

Polysilicon Abbreviation for polycrystalline silicon. An aggregate of small crystalline grains of silicon, each with a different preferred orientation. The grains may vary in dimensions from a few nanometers to a few micrometers.

Polymerase chain reaction (PCR) In biochemistry, it is an amplification process invented in the 1980s for creating billions of identical replicas of a DNA fragment.

Primitive unit The smallest repeating block of a crystal lattice.

p-type doping   *see* doping.

p-n diode   *see* doping.

Quadrature   The situation when two periodic signals of the same frequency, f_L, are out of phase by a quarter of a cycle, or 90°. For example, sine and cosine waveforms are in quadrature (or phase quadrature). One important application is in communications and RF circuits. Separation is possible by heterodyning (multiplication) with another signal of frequency f_R. The amplitudes at the two new frequencies, $(f_R + f_L)$ and $(f_R - f_L)$, are proportional to the amplitude sum and difference, respectively, of the signals in quadrature.

Quality factor   The ratio of the resonant frequency to the bandwidth at −3 dB of a resonant electrical or mechanical system. The sharper the resonance, the higher the quality factor. It is a measure of the frequency stability of oscillators. Physically, it arises from energy loss mechanisms, such as viscous damping or friction at grain boundaries. In an RLC electrical circuit, it is equal to $\sqrt{LC}/R$.

Reduction   Chemical process by which the atoms of an element gain electrons and increase their negative valence. Reduction neutralizes positive ions in an aqueous solution.

Sacrificial etching   A micromachining processing method in which an intermediate layer sandwiched between two layers of a different material is preferentially (sacrificially) etched and selectively removed. Usually, the etch selectivity is high between the intermediate layer and the two sandwich layers. The purpose of the sacrificial layer is to mechanically release one or both of the sandwich layers. Silicon oxide is a commonly used sacrificial layer.

Silicon-fusion bonding   A process to fuse or bond together two silicon substrates. The bond is strong, generally occurring at the molecular level.

Silicon-on-insulator (SOI)   Substrates consisting of a thin layer of silicon dioxide, typically 0.5- to 2-μm-thick, sandwiched between two crystalline silicon layers. The silicon dioxide is known as "buried oxide." One method to fabricate SOI substrates is by silicon-fusion bonding a silicon wafer with a thin layer of silicon dioxide on its surface to a bare silicon wafer. SOI is a well-proven technology for the fabrication of CMOS electronic circuits suitable for high-temperature operation (up to 300° C), as well as for high voltage (> 100 V) and high frequency (< 10 GHz) applications.

Sheet resistance   The resistance of one square of material in units of Ω per square (Ω/ $\square$). It is equal to resistivity divided by the thickness of the material. For thick-film resistors, it is generally implicit that the unit thickness is one mil (25.4 μm).

Solidus temperature   *see* liquidus temperature.

Sound power level (SPL)   Sound pressure, in decibels, measured in reference to a base sound pressure of 20 μPa in air. The reference is usually 1 Pa in water.

Sputtering   A process to deposit a thin film on the surface of a substrate. It involves the removal of material from a target by ion bombardment and subsequent redeposition on the substrate.

SRAM   An acronym for Static Random Access Memory, a type of electronic memory that can be arbitrarily addressed. Unlike EEPROM, it cannot hold the data once electrical power is turned off.

Strain   In mechanics, a deformation produced by stress. In a beam, it is equal to the change in length divided by the original beam length.

Surface micromachining   A class of fabrication processes yielding micromechanical structures that are only a few micrometers thick.

Surface-mount technology (SMT)   An advanced electronic packaging technology in which the type of packages are particularly small so that they can be soldered in high density on the surface of a printed-circuit board.

SVGA   *see* VGA.

SXGA   *see* VGA.

Temperature coefficient of expansion   *see* coefficient of thermal expansion.

Temperature coefficient of resistance (TCR)   The rate of increase in resistance as a function of temperature. In general, $TCR = (\Delta R/R)/\Delta T$, where ($\Delta R/R$) is the fractional change in resistance corresponding to a ΔT change in temperature. It is measured in inverse units of temperature (/°C).

Thermocompression bond   A bonding process involving the melting of an intermediate layer between two substrates pressed against each other. Frequently, the intermediate layer is made of glass.

VGA   An acronym for video graphics adapter, it identifies displays with a resolution of 640 × 480 pixels. SVGA, XGA, and SXGA denote displays with resolutions of 800 × 600, 1024 × 768, and 1280 × 1024 pixels, respectively.

Wavelength division multiplexing (WDM)   A protocol in fiber-optic communication in which digital data is multiplexed on different wavelengths in a single fiber. This effectively increases the bandwidth available in one fiber by increasing the number of channels.

Wet anisotropic etching   Process of etching or removal of material from a silicon substrate with the etch front delineated by crystallographic planes. Potassium hydroxide and tetramethyl ammonium hydroxides are two examples.

Wheatstone bridge   An electrical circuit consisting of four resistors forming two branches electrically connected in parallel, with each branch consisting of two resistors electrically in series. It is useful to measure an imbalance in the values of the four resistors.

Young's modulus   Also known as modulus of elasticity, it is a material constant (in units of pressure) relating mechanical stress to elastic strain. It is indicative of the hardness of the material. For example, diamond has a very high Young's modulus, whereas soft polymers have low values. It often depends on orientation in crystalline materials.

About the Author

Nadim I. Maluf received a B.E. from the American University of Beirut, Lebanon; an M.S. from the California Institute of Technology; and a Ph.D. from Stanford University, all in Electrical Engineering.

Dr. Maluf currently heads the R&D department at Lucas NovaSensor in Fremont, California. He is also a Consulting Professor of Electrical Engineering at Stanford University. Dr. Maluf has over 15 years of industry experience in integrated circuit technology, microelectromechanical systems, sensors and actuators; and their use in medical, automotive, and industrial applications.

Index

A

Accelerometers, 108—19
  applications, 109
  capacitive bulk-
      micromachined, 112—14
  capacitive deep-etched
      micromachined, 118—19
  capacitive surface-
      micromachined, 114—17
  multiaxis, 110
  piezoresistive, 111—12
  range and bandwidth, 110
  shock immunity, 110
  structure, 109, 110
  See also Sensors
Action potential, 239
Actuation
  bimetallic, 240
  electrostatic, 92—93
  magnetic, 94
  method comparison, 95
  methods, 91—95
  piezoelectric, 93
  with shape-memory
      alloys, 94—95
  thermal, 93—94
Actuators, 88, 142—56
  defined, 88
  DMD, 142—47
  grating light valve
      display, 183—86
  high-frequency
      filters, 180—83
  leaf-shaped, 154
  micromachined valves, 147—56
  micromechanical
      resonators, 176—80
  micropumps, 190—92
  optical switches, 187—90
  RF switch over GaAs, 197—98
  thermal, 207
  thermomechanical data
      storage, 192—96
Aluminas, 223
Amorphous silicon
  beam structures from, 21
  defined, 239
  form, 16
  gauge factors, 31
  piezoresistive effect, 31
  See also Silicon

253

Amplification, 239
Angular-rate sensors, 119—33
  from Bosch, 130—33
  from British Aerospace
      Systems, 126—28
  from Daimler Benz, 128—30
  from Delco Electronics, 123—26
  implementations, 122
  "rate grade" performance, 123
  ring shell, 122
  specifications, 123
  tuning fork structure, 122
  vibrating ring, 121, 124—25
  yaw-rate, 123
  See also Sensors
Anisotropic wet etching, 42, 58—62
  corners, 61, 63
  cross-sectional schematic, 58
  defined, 249
  etchants, 58—59
  illustrated, 62, 63
  masking pattern design, 61
  in silicon, 61
  See also Etching
Anodic bonding, 70—71
  defined, 70, 239
  illustrated, 71
  temperature, 70
Anodic stripping voltammetry
    (ASV), 175
ANSYS, 89
Applications, 6—7
  areas of, 5
  future, 161—98
  micromachined valve, 148
  micropump, 190
  photoresist, 51
Aspect ratio-dependent etching
    (ARDE), 68
Atomic-force microscopy (AFM), 193

B

Ball bond, 216
Ball grid array (BGA), 225, 240
Bandwidth
  accelerometer, 110
  defined, 240
Bimetallic actuation, 240
Bonding
  anodic, 70—71, 239
  flip-chip, 219—20
  silicon-fusion, 71—72, 247
  thermosonic gold, 216—17
  wire, 216—18
Borophosphosilicate glass (BPSG), 48
Bosch angular-rate sensor, 130—33
  defined, 130—31
  fabrication process, 132—33
  illustrated, 132
  out-of-phase resonant
      frequency, 131
  sensitivity, 133
  See also Angular-rate sensors
British Aerospace angular-rate
    sensor, 126—28
  closed-loop feedback, 127
  current loop, 126
  defined, 126
  fabrication, 127—28
  illustrated, 127
  specification sheet, 128
  See also Angular-rate sensors
Bulk micromachining, 240

C

Calibration and
    compensation, 211—12
Capacitive bulk-micromachined
    accelerometer, 112—14

Index

Capacitive bulk-micromachined
  accelerometer (continued)
 acceleration rating, 117
 defined, 113
 fabrication process steps, 115
 illustrated, 113
 masking layers, 114
 VTI Hamlin, 112, 113, 114
 See also Accelerometers
Capacitive deep-etched
  micromachined
  accelerometer, 118—19
Capacitive sensing, 91
Capacitive surface-micromachined
  accelerometer, 114—17
 advantage, 117
 ADXL device, 115
 defined, 114
 illustrated, 116
 suspended comb-like structure, 114
 x-axis, 115
 See also Accelerometers
Carbon monoxide gas
  sensor, 136—38
 defined, 136—37
 fabrication process, 138
 illustrated, 137
 MGS1100, 137—38
 operation, 138
 See also Sensors
Cell cultures over
  microelectrodes, 173—75
Ceramic packaging, 223—28
 aluminas, 223
 fabrication process flow, 226
 illustrated, 228
 laminates, 223
 material properties, 225
 of optical MEMS, 227
 types of, 223
 See also Packaging
Chemical sensing, 175—76

Chemical vapor deposition
  (CVD), 46—50
 defined, 46, 241
 deposition of polysilicon, 47—48
 deposition of silicon
  dioxide, 48—49
 deposition of silicon
  nitrides, 49—50
 low pressure (LPCVD), 47, 141, 245
 Plasma-enhanced
  (PECVD), 47, 50, 141
 processes, 47
 thin films deposited by, 46
Chemomechanical polishing
  (CMP), 73, 146
CMOS
 defined, 241
 electronic address, 145
 technology, 9, 146
Coefficient of thermal expansion
  (CTE), 241
Computer-aided design (CAD)
  tools, 88
Conferences, 12—13
Coriolis acceleration, 121
 defined, 120, 241
 illustrated, 121
Corner compensation, 61
Crystalline silicon
 bulk mechanical properties, 19
 properties, 18—19
 tensile yield strength, 18
 wafers, 16, 20
 See also Silicon
"Curie temperature," 33

D

Daimler Benz angular-rate
  sensor, 128—30
 defined, 128
 fabrication process, 129—30

Daimler Benz angular-rate
    sensor (continued)
  illustrated, 129
  tine balancing, 128
  See also Angular-rate sensors
DC glow discharge, 44
Deep reactive ion etching
    (DRIE), 65, 66
  accelerometer, 118—19
  evolution, 66
  limitation, 68
  process characteristic, 69
  profile, 67
  silicon fusion bonding with, 79—81
  See also Etching
Defense Advance Research Program
    Agency (DARPA), 4
Delco angular-rate sensor, 123—26
  defined, 123
  electrodes, 125
  fabrication process, 125—26
  illustrated, 124
  nodes, 125
  specifications, 126
  theory, 124—25
  See also Angular-rate sensors
Diamond, 26—27
Die-attach processes, 212—16
  defined, 213
  organic adhesives, 215—16
  solder, 213—14
  See also Packaging
Diffusion, 241
Digital Micromirror Device
    (DMD), 77, 142—47
  defined, 142
  fabrication process, 145—46, 205
  full-color projection with, 144
  mechanical integrity, 146
  micromirrors, 147
  optical beam steering, 144
  optical switching elements, 142

  package, 211
  pixel illustration, 143
  reliability, 147
Direct wafer bonding. See Silicon
    fusion bonding
DNA, 164
  addressing, 172—73
  capture probes, 172, 173
  fragments, 165, 168, 169
  polymerase, 165
  sequencing, 168, 169
Doping
  defined, 242
  polysilicon, 48
Double-sided lithography, 54—55
  equipment, 55
  example, 54
  See also Lithography
Dry etching, 43, 57, 64—70
Duality, 242
Dupont 951 Green Tape, 223

E

Electrical interconnects, 216—20
  flip-chip bonding, 219—20
  wire bonding, 216—18
  See also Interconnects
Electrochemical etching, 62—64
  crystalline silicon island, 66
  defined, 62—64
  illustrated, 65
  in original implementation, 64
  See also Etching
Electromagnetic interference
    (EMI), 205, 242
Electrophoresis, 168—71
  defined, 243
  demonstration, 170—71
  fluid injection, 170
  illustrated, 170

miniaturization, 168
research activities, 169
separation step, 170
Electroplating, 75
Electrostatic actuation, 92—93
Electrostatic comb filters, 179
Epitaxy, 43—44
  defined, 43, 243
  growth, 43—44
  use of, 43, 44
Etching, 55—70
  aspect ratio-dependent (ARDE), 68
  deep reactive ion (DRIE), 66
  dry, 43, 57
  electrochemical, 62—64
  plasma-phase, 64—70
  process, 55
  reactive ion (RIE), 43, 66
  sacrificial, 247
  silicon, 56
  thin films, 56, 57
  trench profiles, 68
  wet, 43, 58—62
European Microsystem Technology
      On-line (EMSTO), 11
Eutectic point, 243
Evaporation, 45—46
  defined, 45
  directional deposition process, 46
  target heating and, 45—46

F

Fabrication, 6
  Bosch angular-rate
      sensor, 132—33
  British Aerospace angular-rate
      sensor, 127—28
  capacitive bulk-micromachined
      accelerometer, 115
  carbon monoxide gas sensor, 138
  ceramic packaging, 226
  Daimler Benz angular-rate
      sensor, 129—30
  Delco angular-rate
      sensor, 125—26
  DMD, 145—46, 205
  etching process for, 55
  grating light valve (GLV), 186
  Hewlett-Packard micromachined
      valve, 155—56
  infrared imager, 135—36
  low-cost batch, 57
  microelectrode arrays, 171—72
  micromachined microphone, 141
  micropump, 192, 193
  molded plastic
      packaging, 230—35
  nozzle, 95
  optical switches, 188—89
  piezoresistive accelerometer, 112
  pressure sensor, 101
  processes, 25
  Redwood Microsystems
      micromachined valve, 151
  RF switch over GaAs, 197—98
  SFB-DRIE, 80
  SOI high-temperature pressure
      sensor, 106
  thermomechanical data
      storage, 195—96
  TiNi Alloy micromachined
      valve, 153—54
Faradaic current, 176
Flip-chip bonding, 219—20, 225
  advantages, 219
  defined, 219
  fabrication steps, 219—20
  illustrated, 220
  incompatibility, 220
  See also Electrical interconnects;
      Wire bonding
Fluid nozzles, 95—97

Fluorinert, 149—50, 151
Force balancing, 117
Foundry, 243
Frequency degeneracy, 243

G

Gallium arsenide (GaAs), 27
Genetic diagnostics, 165
Gold metalization, 104
Grating light valve (GLV), 183—86
    advantage, 185—86
    defined, 183
    fabrication, 186
    full color display, 185
    gray shade support, 185
    operating principle, 184
    optical projection system, 184
    pixels, 184
Grinding, 72—73
Gyroscopes, 119—33
    mechanical, 120
    precision, 119
    ring-laser, 119

H

Helmholz cavity, 243
Hermetic packaging, 210—11
    defined, 210
    interior, 211
    material, 210
    See also Packaging
Hewlett-Packard micromachined valve, 154—56
    cross-sectional illustration, 156
    defined, 154—55
    fabrication, 155—56
    illustrated, 155
    See also Micromachined valves

High-frequency filters, 180—83
    defined, 180
    effect, 180—81
    example, 182
    photograph, 183
    traveling waves and, 182
    See also Actuators
High-temperature pressure sensors, 104—5
Hinge mechanisms, 162—63
    defined, 162
    demonstrations, 163
    illustrated, 162
    structure, 162—63
    See also Passive structures
Hole, 244
Honeywell AWM sensors, 106, 107
Hybridization, 244

I

Impedance, 244
Infrared imager, 134—36
    defined, 134
    fabrication, 135
    illustrated, 134
    read-out electronics, 136
    See also Sensors
Inkjet print nozzles, 97—98
Insertion loss
    defined, 244
    optical switch, 189—90
Interconnects, 216—22
    electrical, 216—20
    microfluidic, 220—22
    See also Packaging
International Society for Optical Engineering (SPIE), 13
Ion implantation, 244
Isotropic wet etching, 57, 58

Index

J

Journal of Micromechanical Systems, 11
Journal of Micromechanics and Microengineering, 11
Journals, 11—12

K

Kovar, 225, 227

L

Laser ablation, 212
Liquidus temperature, 244
Lithography
    contact, 51—52
    defined, 42, 244
    double-sided, 54—55
    large field of view, 55
    photoresist, 51
    projection, 52, 53
    proximity, 52
    resolution, 53
    steps, 51—55
    thick resist, 53
    topographical height variations, 54
Lorentz force, 245
Low pressure CVD (LPCVD), 47, 141, 245

M

Magnetic actuation, 94
Manufacturing volume, 9
Markets, 6—7
    analysis/forecast, 7
    end, 8
    estimate, 6
Mass flow sensors, 105—8
    Honeywell AWM, 106, 107
    illustrated, 108
    See also Sensors
Materials, 15—37
    diamond, 26—27
    gallium arsenide (GaAs), 27
    glass substrates, 26
    polymers, 25
    properties, 17
    properties/physical effects, 28—37
    quartz, 26, 35
    selectively removing, 43
    shape-memory alloys, 27—28
    silicon, 16—23
    silicon carbide, 25, 26—27
    silicon oxide and nitride, 23
    thin metal films, 23—25
Media isolation, 210
MEMCAD, 89
MEMS
    application areas, 5
    categories, 88
    components, 3—4
    conferences, 12—13
    defined, 3, 245
    journals, 11—12
    materials for, 15—37
    packaging, 201—35
    product features, 4
    psychological barriers, 10
    solutions, 8
    standards, 9—10
    structures, 87—156
    technology, 4, 9, 10
    use decision, 7—9
    Web sites, 10—11
MEMS Clearinghouse, 11
Metal packaging, 228—30
    defined, 229
    hermetic, 229

Metal packaging (continued)
    illustrated, 230, 231
    solutions for harsh
        environments, 229—30
    See also Packaging
Microelectrode arrays, 171—76
    cell cultures over, 173—75
    chemical sensing of trace metals
        with, 175—76
    cross-section, 172
    defined, 171
    DNA addressing with, 172—73
    fabrication, 171—72
    photograph, 177
    research, 171
    See also Sensors
Microelectromechanical systems.
        See MEMS
Microfabrication process, 75
Microfluidic interconnects, 220—22
    demonstration, 221—22
    illustrated, 221
    requirement, 221
    See also Interconnects; Packaging
Micromachined components, 4, 6
Micromachine Devices, 12
Micromachined microphone, 138—41
    corrugated circular diaphragm, 140
    defined, 139
    electret, 242
    fabrication, 141
    illustrated, 140
    Knowles, 140, 141
    sensitivity, 139, 141
    technical characteristics, 139
    See also Sensors
Micromachined valves, 147—56
    field of, 147
    Hewlett-Packard, 154—56
    potential applications, 148
    Redwood Microsystems, 148—51
    TiNi Alloy, 152—54

See also Actuators
Micromachining
    bulk, 240
    conferences, 12—13
    defined, 6, 245
    journals, 11—12
    polysilicon surface, 77—79
    process, 42
    process flow illustration, 43
    silicon, 42
Micromechanical resonators, 176—80
    illustrated, 180
    properties, 179
Micropumps, 190—92
    applications, 190
    fabrication, 192, 193
    illustrated, 191
    pump rate, 192
    stand-alone, 190
    structure, 190—91
Microsystems, 4, 5
Microsystems technology (MST)
    defined, 3
    See also MEMS
MicroTotalAnalysis Systems
        (MTAS), 13
Miniature biochemical reaction
        chambers, 163—68
Molded plastic packaging, 230—35
    approaches, 231
    defined, 230—31
    fabrication, 232—33
    illustrated, 233, 234
    postmolding, 231, 232
    premolding, 231, 232, 233
    schematic, 232
    standard, for integrated
        circuits, 234
    See also Packaging
Molding, 75
MST News, 12
Multichip modules (MCM), 245

Index 261

N

Nanogen electronic addressing, 174
Nitinol, 28
Noise, 245
Nozzles, 95—98
    circular, 96
    fabrication of, 95
    fluid, 95—97
    illustrated, 96, 97
    inkjet print, 97—98
    square, 96

O

Optical add/drop multiplexers (OADM), 187
Optical switches, 187—90
    characteristics, 187
    defined, 187
    demonstration, 188
    fabrication, 188—89
    illustrated, 189
    insertion loss, 189—90
    See also Actuators
Organic adhesives, 215—16
Organization, this book, xviii—xix
Oxidation, 44, 246

P

Packaging, 201—35
    calibration and compensation, 211—12
    categories, 222
    ceramic, 223—28
    cost, 222—23
    defined, 201
    design factors, 202—3
    die-attach processes, 212—16
    DMD, 211
    evolution, 202
    field breadth, 202
    first-level, 222
    hermetic, 210—11
    media isolation, 210
    metal, 228—30
    molded plastic, 230—35
    pressure sensor, 202
    process flow illustration, 203
    protective coatings, 208—10
    requirements, 224
    solution types, 222—35
    standards, 202
    stress isolation, 207—8
    thermal management, 205—7
    wafer dicing concerns, 204—5
    wafer/wafer-stack thickness, 204
    wiring and interconnects, 216—22
Passive structures, 88, 95—98
    defined, 88
    fluid nozzles, 95—97
    hinge mechanisms, 162—63
    inkjet print nozzles, 97—98
Peltier devices, 35—36
Phosphosilicate glass (PSG), 48
Photoresist
    application of, 51
    defined, 51
    positive, 51
    spin coating, 54
    thick, 53
Piezoelectric actuation, 93
Piezoelectricity, 31—35
    coefficients, 33, 34
    defined, 31—32, 246
    illustration, 33, 34
    physical origin of, 32
    quartz and, 35
    as sensing method, 90—91

Piezoelectricity (continued)
　　See also Materials; Piezoresistivity;
　　　　Thermoelectricity
Piezoresistive accelerometer, 111—12
　　Endevco, 112
　　fabrication, 112
　　illustrated, 111
　　See also Accelerometers
Piezoresistive gauge, 101
Piezoresistivity, 29—31
　　cause of, 29
　　coefficients, 30, 31
　　defined, 29, 246
　　p-type, 30
　　resistor direction, 29—30
　　as sensing method, 90—91
　　See also Materials; Piezoelectricity;
　　　　Thermoelectricity
Planar RF, 44—45
Plasma-enhanced CVD
　　　　(PECVD), 47, 50, 141
Plasma-phase etching, 64—70
　　ARDE, 68
　　defined, 246
　　DRIE, 66, 67
　　leading developers of, 64
　　principle of, 65—66
　　RIE, 66
　　uses, 64—65
　　See also Etching
Polishing, 72—73
Polymerase chain reactions
　　　　(PCR), 23
　　chamber illustration, 167
　　cycles, 167
　　defined, 246
　　demonstrations, 165—66
　　fabrication, 166—67
　　illustrated, 166
Polymers, 25
Polysilicon
　　beam structures from, 21

defined, 246
deposition of, 47—48
doping, 48
electrical properties, 20
films, 48
form, 16
gauge factors, 31
mechanical properties, 21
piezoresistive effect, 31
surface micromachining, 77—79
uses, 20—21
See also Silicon
Pressure sensors, 99—104
　　fabrication, 101
　　packaging, 202
　　piezoresistive, 99
　　schematic illustration, 100
　　sensitivity, 100
　　silicon-fusion bonded, 102, 103
　　silicon-on-insulator, 104, 105
　　structure, 99
　　See also Sensors
Projection lithography
　　illustrated, 52
　　resolution, 53
　　superiority, 53
　　See also Lithography
Protective coatings, 208—10
　　for extreme
　　　　environments, 209—10
　　material properties, 209
　　silicon carbide, 209
Proximity lithography, 52
Pyrex glass wafers, 102

Q

Quadrature, 247
Quality factor, 247
Quartz
　　as piezoelectric material, 35
　　tuning forks, 122

Index

R

Radiation sensors, 134—36
Reactive ion etching (RIE), 66
Redwood Microsystems
    micromachined
    valve, 148—51
  defined, 148—49
  fabrication steps, 151
  Fluorinert perfluorocarbon
    liquids, 149—50
  illustrated, 149
  NO-1500 Fluistor, 151
  operating mechanism
    illustration, 150
  See also Micromachined valves
RF switch over GaAs, 197—98
  defined, 197
  fabrication, 197—98
  illustrated, 196
Room-temperature vulcanizing
    (RTV) silicons, 215—16

S

Sacrificial etching, 247
SCREAM, 81—82
  defined, 77
  process, 81
  steps illustration, 82
Seebeck coefficients, 36—37
Seebeck effect, 36
Sensing
  capacitive, 91
  with electromagnetic signals, 91
  historical accounts, 90
  method comparison, 92
  methods, 90—91
  objective, 90
  with piezoelectricity, 90—91
  with piezoresistivity, 90—91

Sensors, 88, 99—141
  accelerometer, 108—19
  angular-rate, 119—33
  calibration, 211—12
  carbon monoxide gas, 136—38
  defined, 88
  electrophoresis on a chip, 168—71
  high-temperature
    pressure, 104—5
  magnetoresistive, 91
  mass flow, 105—8
  microelectrode arrays, 171—76
  micromachined
    microphone, 138—41
  miniature biochemical reaction
    chambers, 163—68
  performance characteristics, 211
  pressure, 99—104
  radiation, 134—36
  yaw-rate, 91
Sensors and Actuators, 11
Sensors Magazine, 12
SFB-DRIE, 79—81
S-gun, 45
Shape-memory
    alloys, 27—28, 94—95
Sheet resistance, 248
Silicon, 16—23
  amorphous, 16
  crystalline, 16, 18—20
  cut plane, 18
  diamond lattice structure, 17
  etching, 56
  forms, 16
  interactions, 22
  mechanical integrity, 22
  micromachining, 42
  micropump, 191
  nozzle, 96
  optical reflectivity of, 22
  polysilicon, 16, 20—21, 31,
    47—48, 77—79

Silicon (continued)
  principal axes, 17
  RTV, 215—16
  thermal conductor, 21
Silicon carbide, 25, 209
Silicon dioxide, 44
  deposition of, 48—49
  deposition rates, 49
Silicon-fusion bonding, 71—72
  defined, 71, 247
  with DRIE, 79—81
  "hydration" step, 72
  mechanism, 71—72
  pressure sensors, 102, 103
  See also Bonding
Silicon material system, 16—25
  polymers, 25
  silicon, 16—23
  silicon oxide and nitride, 23
  thin metal films, 23—25
  See also Materials
Silicon nitride, 23
  deposition of, 49—50
  films, 49
Silicon-on-insulator (SOI)
    wafers, 69, 104, 105,
247—48
Silicon-on-sapphire (SOS) wafers, 44
Silicon oxide, 23
Single Crystal Reactive Etching And
    Metalization. See SCREAM
Solders, 213—15
  choice of, 213
  phase diagram, 215
  properties, 214
  silicon/glass and, 214
  See also Die-attach processes
Sol-gel deposition methods, 74
Sound power level (SPL), 248
Spin-on methods, 50—51
  defined, 50
  materials, 50

Sputter deposition, 44—45
  of aluminum, 82
  defined, 44
  deposited film, 45
  use of, 45
Sputtering, 248
Standards, 9—10
Static-random-access-memory
    (SRAM) cells, 143
Stitch bond, 217
Stress isolation, 207—8
Surface micromachining
  defined, 248
  polysilicon, 77—79
  steps illustration, 78
Surface mount technologies
    (SMT), 202

T

Temperature coefficient of
    expansion, 248
Temperature coefficient of resistance
    (TCR), 31, 90, 103, 106, 136,
    248—49
Thermal actuation, 93—94
  approaches, 93—94
  power consumption, 93
  See also Actuation
Thermal actuators, 207
Thermal management, 205—7
  demands on, 205
  levels, 205—6
  at package level, 207
Thermocompression bond, 249
Thermoelectricity, 35—37
  defined, 35
  See also Materials; Piezoelectricity;
    Piezoresistivity
Thermomechanical data
    storage, 192—96

Index

cantilevers, 194
defined, 193
demonstration, 193
fabrication, 195—96
Thermosonic gold bonding, 216—17
Thin metal films, 23—25
   choice of, 23
   depositing, 23
   etching, 56, 57
   metals list, 24
   use of, 23—25
TiNi Alloy micromachined
      valve, 152—54
   assembly, 152—53
   defined, 152
   fabrication, 153—54
   illustrated, 152
   performance advantage, 154
   See also Micromachined valves
Toolbox, 6
   advanced tools, 70—75
   anodic bonding, 70—71
   basic tools, 42
   chemical vapor deposition
      (CVD), 46—50
   electroplating and molding, 75
   epitaxy, 43—44
   etching, 43, 55—70
   evaporation, 45—46
   grinding, polishing, and
      CMP, 72—73
   lithography, 42, 51—55
   oxidation, 44
   silicon-fusion bonding, 71—72
   sol-gel deposition methods, 74
   spin-on methods, 50—51
   sputter deposition, 44—45
   tool combinations, 75—82

V

Volume manufacturing, 9

W

Wafers
   crystalline silicon, 16, 20
   dicing concerns, 204—5
   Pyrex glass, 102
   SOI, 69, 104, 105, 247—48
   SOS, 44
   thickness, 204
   See also Packaging
Wavelength division multiplexing
      (WDM), 187, 249
Web sites, 10—11
Wet etching, 43, 58—62
   anisotropic, 58—62, 249
   cross-sectional schematic, 58
   isotropic, 58
   See also Etching
Wheatstone bridge, 249
Wire bonding, 216—18
   to aluminum bond pads, 217
   illustration, 218
   limitations, 218
   thermosonic gold
      bonding, 216—17
   See also Electrical interconnects;
      Flip-chip bonding
Wiring, 216—22

Y

Young's modulus, 249

Recent Titles in the Artech House Microelectromechanical Systems (MEMS) Series

An Introduction to Microelectromechanical Systems Engineering, Nadim Maluf

Introduction to Microelectromechanical (MEM) Microwave Systems, Hector J. De Los Santos

For further information on these and other Artech House titles, including previously considered out-of-print books now available through our In-Print-Forever® (IPF®) program, contact:

Artech House
685 Canton Street
Norwood, MA 02062
Phone: 781-769-9750
Fax: 781-769-6334
e-mail: artech@artechhouse.com

Artech House
46 Gillingham Street
London SW1V 1AH UK
Phone: +44 (0)20 7596-8750
Fax: +44 (0)20 7630-0166
e-mail: artech-uk@artechhouse.com

Find us on the World Wide Web at:
www.artechhouse.com